Longman

Oliver & Boyd

First published 1984
ISBN 0 582 17235 7

Printed in Great Britain by
Longman Group – Resources Unit, York.

PREFACE

Aspects of Geology takes a close look at six topics which are crucial to an understanding of the nature and aims of modern geological study. Finding out about the history of the earth, the way it has evolved and the processes that are constantly at work in itself makes fascinating reading. The aim of the book, however, is also to show the vital importance of a range of geological disciplines in the modern world, not least in the search for mineral wealth and fossil fuels such as coal and oil. It is these practical demands on the geologist, reflected in the case studies discussed, that link together the often very diverse interests to be found within the field of geology.

The book has been written with the needs of the A level student in mind, but many amateur enthusiasts and undergraduates needing a readable introduction to the subject will find the material of interest. Complete coverage is not attempted, rather the general principles are introduced and applied in depth to a number of case studies that are relevant to the applications of geological study today.

Exercises of varying complexity are included and in some instances these constitute the major part of a case study. It is important that these be attempted if the reader is to benefit fully from the material provided.

CONTENTS

Palaeoecology

Contents

Acknowledgements

The authors are most grateful to the following for their considerable help in the preparation of this Collection: Mr K B Sale for the Scarborough beach data; Mr D G Elford for his comments on the Frodingham Ironstone section; Drs F Spode and M Romano for their critical reading of the first draft and help with sources of information; Mrs B G Ross and Mrs M Whitehead for help with the typing.

The authors accept full responsibility for any remaining shortcomings and errors.

We are also grateful for permission to reproduce the following photographs:
Fig. 1.1 from Raup D M and Stanley S M, *Principles of Palaeontology*, W H Freeman and Co; Figs 5.2 and 5.3, the British Steel Corporation, Scunthorpe Division; Figs 6.1 and 6.2 from Ennion E A R and Tinbergen N, *Tracks*, Clarendon Press, 1967; Fig. 6.5 from Bassett M G and Owens R M, *Fossil Tracks and trails*, National Museum of Wales.

The following figures have been based on the sources listed.

Fig. 1.2, *Geology Journal* Vol 69, p. 159 by Logan B W; Fig. 2.2b Miles P M and H B, *Seashore Ecology*; Figs. 2.4, 2.7 and 4.3 Raup D M and Stanley S M, *Principles of Palaeontology*; Fig. 2.5 Laporte L F, *Ancient Environments*; Fig. 4.1 Kirkaldy J F, *The Study of Fossils*; Fig. 4.7 Ford T D, *Limestone and Caves of the Peak District*; Figs. 5.5 and A.1 McKerrow W S, *Ecology of Fossils*; Fig. 6.6 *Proceedings of the Yorkshire Geological Society*, Vol 41 part 4 1978; Fig. 7.3 Smith F D, *Micropalaeontology 1*; Fig. 7.4 West R G, *Pleistocene Geology and Biology*.

Foreword

The purpose of this Unit is to examine the ecological relationships of a variety of organisms, living and fossilized. No attempt has been made to deal with the morphology or biological classification of the organisms, since it is assumed that the reader will already be familiar with these aspects, at least in respect of the major groups of invertebrate animals.

The case studies in Sections 4 to 7 are designed to demonstrate the use of palaeoecological principles in a wide variety of ancient environments of several different geological ages. Where appropriate, the Principle of Uniformitarianism provides the underlying philosophy, although it is applied with varying degrees of certainty according to the age and nature of the deposits and their included fossils.

Work suggestions have been printed in italic type and it is intended that you work out the answers as you go along. Some sections depend rather heavily on such exercises being done, so it is important that you do not miss them out!

1
Introduction

A walk along a beach or through a wood will reveal a great variety of life: plants, insects even vertebrates. After all, man is a vertebrate. All these different organisms depend upon each other to a certain extent, either as sources of food, or for protection. For example, some animals use plants for protection, in return for which the animals will give back nutrients to the soil.

Ecology is the study of the relationship of these plants and animals to each other and to their environment. On the earth today there are a tremendous number of different environments containing organisms which relate to each other in a variety of ways. These include all the different areas beneath the sea as well as the more familiar environments on land. Most of the relationships between organisms can be studied directly by the biologist. He can also observe variations in the delicate balance that exists within these habitats.

In the past there must also have been a great variety of such interrelationships but the geologist is not as fortunate as the biologist, in that he cannot actually study them in operation. He has to make do with the fossil evidence and try and reconstruct what the environment must have been like. But to what extent can the geologist rely on fossil evidence? There are many horizons in the stratigraphic record that lack fossils completely, but does this necessarily mean that there was no life in the area at the time, or is it merely because the life-forms that did exist did not become fossilized for a variety of reasons? Even a rock type that reveals a rich variety of flora and fauna such as the Silurian Wenlock Limestone can be misleading, as some of the life forms that existed at the time, such as soft bodied organisms, may be missing. Conversely, they may even have fossils in them that have been transported after death. Some of these problems of fossilization and the partial nature of the fossil record are discussed further in Section 3 of this Unit.

Uniformitarianism

When reconstructing past environments, the geologist relies to a great extent on comparing them with what he believes are similar environments today. This principle of comparison is known as Uniformitarianism. It is a basic concept in geology and it dates back to the time of the Scottish geologist James Hutton (1726-97) who was one of the first to realise that what was happening on the earth's surface today, such as erosion and deposition, must also have been going on in the past. He came to this conclusion after studying sedimentary rocks near his home in Edinburgh.

The principle of Uniformitarianism has been encapsulated in the phrase, "The present is the key to the past". But to what extent is it true that processes on the earth's surface today are typical of what has happened in the past? Generally, it seems as if there is close agreement and the study of modern environments provides a good basis for the reconstruction of the past, although the level of comparability between modern and ancient decreases with time.

With the more recent geological periods it is relatively easy to find fossils that are closely related to organisms living today. It is possible to state with some certainty that a fossil of *Ostrea bellovacina* found in the Tertiary would have had a very similar mode of life to a modern *Ostrea* (oyster). However, what happens if the fossil is from an extinct group? By definition there are no living members with which comparison can be made. Here, there is plenty of scope for the geological detective.

The trilobites are an extinct group but we have quite an accurate idea of their mode of life. The type of rocks in which they are found and the other fossils associated with them provide valuable clues. Trilobites are found with brachiopods and corals, which we know by comparison with modern groups live in the sea; they are not found with land dwelling organisms. We can therefore say that trilobites were marine creatures.

Further evidence comes from studies of the morphology of trilobites. Trilobites have hard exoskeletons on the upper sides of their bodies. We know from exceptionally well preserved ones that they had legs on their undersides. Where eyes are present, they are usually on the tops of their heads and would have given the trilobite a clear field of vision into the surrounding water. Taken together, these observations indicate a mode of life on the sea bed, where the animal probably crawled around in search of food. Some trilobites lack eyes and so it seems that they must have inhabited areas where eyes were not needed. Perhaps they lived in dark burrows in the sediment or in deep water through which the light did not penetrate.

Another line of approach with extinct groups is to compare them with their closest living relatives. A very accurate idea of the mode of life of the extinct cephalopod groups, the goniatites and ammonites, can be obtained by comparing them with a living cephalopod group, the nautiloids.

With other groups comparisons are not quite so easy. The exact mode of life of dendroid graptolites is still not quite clear as they do not have any close living relatives and their morphologies do not readily suggest a particular mode of life.

Caution!

Even when living relatives of fossil organisms are known, we have to be careful not to misapply the principle of Uniformitarianism. For example, stromatolites are some of the earliest fossils found in the Precambrian. They have been recorded from rocks such as the 1,900 million year old Gunflint Chert on the shores of Lake Superior, or the even older Bulawayo Limestone of Zimbabwe which may be more than 2,700 million years old. It is believed that they have been formed by primitive organisms related to the algae known as cyanophytes or "blue-greens". Further examples of stromatolites are found throughout the fossil record and the layered structure that they all show closely resembles modern forms.

Modern stromatolites are found in rather restricted intertidal areas in the Bahamas and Western Australia. They occur as mounds of around a metre in height built up of alternating layers of sand impregnating the algal mat and carbonates precipitated by algae. The algae are mainly the very primitive "blue-greens" together with some green algae (Fig. 1.1).

Fig. 1.1 Recent stromatolites in Shark Bay, Australia, orientated parallel to the direction of the wave surge.

Ancient stromatolites in the Precambrian Great Slave Lake area, Canada. Elongated again, parallel to the ancient current direction.

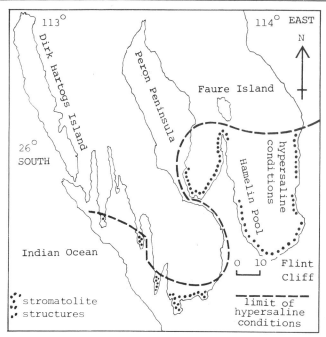

Fig. 1.2 Shark Bay, Western Australia.

In the parts of Shark Bay, Western Australia (Fig. 1.2) where the stromatolite mounds are found growing today, a sandbar has largely cut off direct access to the sea. The hot climate causes evaporation of the water behind the sandbar making it very saline. This does not hamper the growth of the stromatolites but molluscs and other creatures which would normally feed on the "blue-greens" can not survive. Stromatolites are only found today in areas such as Shark Bay where they can grow unchecked because predators are kept out by hostile conditions.

A strict application of the theory of Uniformitarianism would have it that stromatolites would always have been found in these rather unpleasant, hot, saline conditions. Back in the Precambrian, however, when the stromatolites were the most advanced life form, they would have had a much wider distribution, as they could have flourished uncropped in a variety of different environments.

This case illustrates that there has to be a certain amount of caution in relating modern environments and conditions to the past. The changing variety and abundance of life forms causes different pressures. In the past, stromatolites survived unchecked in many different environments, but today competition and predators have forced them into the rather unfavourable habitats in which we now find them.

Nevertheless it is still very important to be familiar with modern environments and processes in order to be able to fully appreciate the situations that existed in the past.

The variety of modern environments is far too great to be able to describe them all in this Unit and so we have selected one particular example to look at in detail. At first glance this may seem to be a simple, familiar environment, but on closer examination, the modern beach turns out to be a complex, delicately balanced system.

2
Modern Environments

A Beach Study

Those who have studied biology at school will be used to the idea of investigating selected habitats from an ecological point of view. Woodland areas, rivers, ponds and sea shores form accessible areas of study. The following example, Scarborough beach in Yorkshire, is a common type of field exercise which students are frequently asked to undertake. Even the casual observer will soon notice that there is an astounding variety of plant and animal life on the sea-shore.

During the course of one hour's collecting between the high and low water marks, it is possible to find about one hundred different organisms on a beach as diverse as Scarborough. Some of these seem to be ubiquitous in their distribution, whilst others are clearly more localised, so the first job the biologist tackles is to plot the distribution and look for patterns.

In the case of Black Rocks at Scarborough, the beach is backed by a cliff some 25 m high. Below this, the upper shore is cluttered with large fallen sandstone rocks. Below this again, there is a gently sloping expanse of rocky platforms extending from the upper shore to the lower, a total distance of some hundreds of metres. In addition, the platform is dissected by numerous gullies running parallel and at right angles to the incoming tide (Fig. 2.1). The solid geology consists of calcareous shales and sandstones.

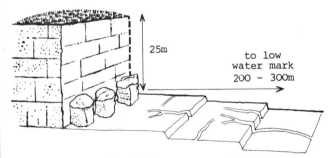

Fig. 2.1 Sketch of the coast at Black Rocks, Scarborough.

In the first instance, reconnaissance sampling is carried out by making a transect at right angles to the falling tide. Sampling frequency is determined by stopping at regular intervals along a rope marked in metres and stretched from a fixed point near the high water mark. The actual sampling is done by the standard biological technique of dropping a quadrat. This is a wooden frame, 50 cm square, subdivided into 25 smaller

PLANT SPECIES	STATION NUMBER													
	UPPER SHORE					MID SHORE							LOW SHORE	
	1	2	3	4	5	6	7	8	9	10	11	12	13	14
Pelvetia	x	x												
Fucus spiralis		x	x	x										
F. vesiculosus		x	x	x	x									
F. serratus									x	x	x			
Corallina				x	x	x	x	x	x	x				
Laurencia			x	x	x	x	x	x	x	x	x			
Laminaria												x	x	x

Fig. 2.2a Distribution of seaweeds along a beach transect at Black Rocks.

squares. Organisms occurring within the area of the quadrat are counted, by a variety of techniques. In Fig. 2.2a seaweed species are shown simply by whether or not they are present in each of the sample squares along the line. Even this simple method demonstrates that plants such as *Pelvetia* evidently require specialised conditions, whereas *Laurencia* is more tolerant. The plants are shown in Fig. 2.2b.

Fig. 2.3 shows the extra detail which may be gained by counting the frequency of occurrence of each of several animal species in the small squares of the quadrat. In this case, a figure of 25 indicates that a species is so common that it occurs at least once in each small square. A similar pattern emerges of restricted habitats for some organisms and widespread ones for others.

Having gathered the initial data, the biologist considers critically any sampling errors, then attempts to set up hypotheses to explain the pattern, hoping where possible to test his ideas by further observation. The following list, which is not exhaustive, is an indication of the complexities of the apparently "simple" beach environment. Any, or all of these factors may be significant in determining the survival or otherwise of a given species.

1. Period of exposure to desiccation at low tide
2. Range of temperature variation of the environment
3. Intensity of light
4. Aspect (north-facing, south facing etc.)
5. Predation by land, sea or aerial predators
6. Wave buffeting
7. Nature of rock surface, type of rock, degree of attachment possible
8. pH and salinity of water
9. Length of time between tides for feeding
10. Pollution

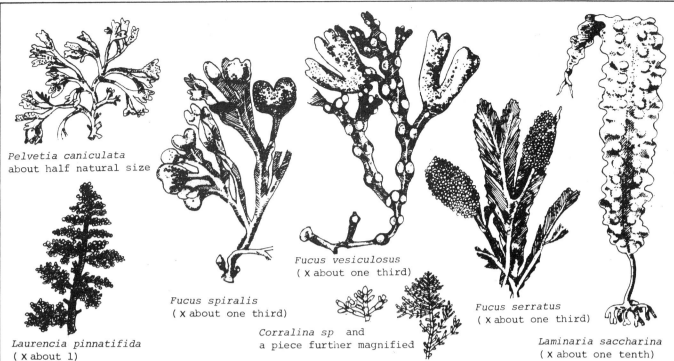

Pelvetia caniculata
about half natural size

Fucus spiralis
(x about one third)

Fucus vesiculosus
(x about one third)

Fucus serratus
(x about one third)

Laurencia pinnatifida
(x about 1)

Corralina sp and
a piece further magnified

Laminaria saccharina
(x about one tenth)

Fig. 2.2b Illustration of seaweeds to be found at Black Rocks.

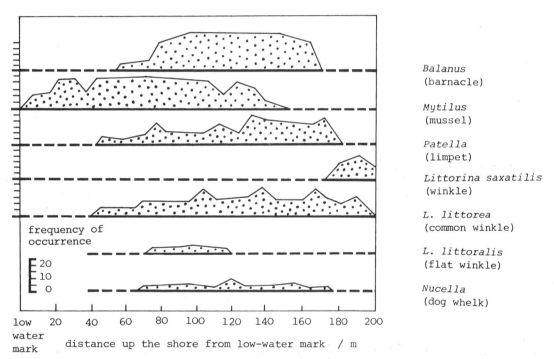

Balanus
(barnacle)

Mytilus
(mussel)

Patella
(limpet)

Littorina saxatilis
(winkle)

L. littorea
(common winkle)

L. littoralis
(flat winkle)

Nucella
(dog whelk)

frequency of
occurrence

20
10
0

low
water
mark

20   40   60   80   100   120   140   160   180   200

distance up the shore from low-water mark / m

Fig. 2.3 Frequency of occurrence of some animal species along a beach transect at Black Rocks.

*How many of these factors would you be able
to allow for in examining a "fossil" beach
deposit?*

Even with the benefit of being able to watch
the living organisms and take some of the
measurements mentioned above, it is not always possible for the biologist to be sure
that he has explained all the distributions.
How much more difficult it is for the geologist, and how tempting to make up a good
story, knowing that it cannot really be
checked since the organisms involved are all
thoroughly dead!

Another line of approach is to examine one
part of the beach, say a rock pool, and to
try to determine the relationship between the

various organisms within it. Such a single
unit, where each plant or animal interrelates
with one or more others is termed an
ecosystem, although one must not forget the
changes in the system at a different state of
the tide, or when a passing seagull drops in
for a feed! Each organism in such a balanced
environment occupies its own microhabitat,
or ecological niche, which it exploits to the
full, not always in such competition with
other species as might be thought. Thus the
winkle (*Littorina littorea*) and the sea
anemone (*Actinia equina*) may co-exist in the
same pool, since the winkle thrives by
grazing on seaweed whilst the sea anemone
catches small crustacea and carrion on its
tentacles, remaining more or less fixed to
the one spot.

Life Habits of Marine Invertebrates

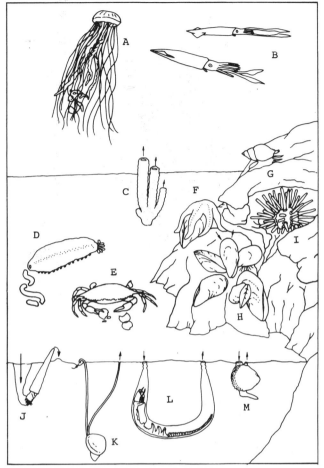

Fig. 2.4 Life habits of some common marine invertebrates.

A composite diagram of the variety of life habits of some common marine invertebrates is given in Fig. 2.4. The organisms are pictured living in close proximity, but the accompanying table (Table 1) shows how little competition there is for the same food, because of the subtle variations in the food source which each organism is exploiting.

An explanation of the terms used in the table is as follows:

Pelagic   — living in the water mass: subdivided into:
              planktonic — floating
              nektonic   — swimming
Benthonic — living on or in the sea bed (substrate) subdivided into:
              epifaunal — living on the substrate
              infaunal — living in the substrate
Active   — able to move by swimming, crawling etc.
Sessile   — remaining more or less fixed in one place
Carnivore/scavenger — catches animal food or searches for dead matter
Suspension feeder — filters minute food particles from sea water
Grazer   — herbivore, usually eating algae
Deposit feeder — ingests sediment and absorbs organic matter from it

TABLE 1 The life habits of the animals shown in Fig. 2.4

NAME OF ORGANISM	PELAGIC		BENTHONIC		MOVEMENT		FEEDING PROCESS			
	planktonic	nektonic	epifaunal	infaunal	active	sessile	carnivore scavenger	suspension feeder	grazer	deposit feeder
A Jellyfish	✓				✓		✓			
B Squid		✓			✓		✓			
C Sponge			✓			✓		✓		
D Sea cucumber			✓		✓					✓
E Crab			✓		✓		✓			
F Starfish			✓		✓		✓			
G Snail			✓		✓				✓	
H Mussel			✓			✓		✓		
I Sea urchin			✓		✓				✓	
J Trumpet worm				✓	✓					✓
K Macomid bivalve				✓	✓					✓
L Chaetopterid worm				✓		almost sessile		✓		
M Cockle				✓	✓			✓		

The diagram is designed to demonstrate a range of ecological niches and does not purport to show the balance between the numbers of <u>individuals</u> following a particular life-style. In practice one can often define a predator/prey ratio, which remains fairly constant for a particular ecosystem. An everyday example would be the ratio between foxes and rabbits inhabiting a patch of heathland. In normal circumstances, there will always be far fewer foxes than rabbits.

Some palaeoecologists have used modern predator/prey ratios to try to reconstruct ancient environments where only hard-bodied predatory animals have left any record of their presence. It is then argued that other, presumably soft-bodied, organisms must have been present for the predators to have had a food supply. The "most likely" prey is then inserted into the reconstruction. This approach is clearly hazardous but is not infrequently attempted.

A General Survey of Modern Environments

Land environments

The habitats of land plants and animals are naturally better known to most of us than are the marine ones. Nevertheless, land areas form only about 29% of the earth's surface. Many regions are being actively eroded and there is little chance of organic remains being preserved unless they are as minute and indestructible as spores or pollen. These may well survive the rigours of erosion and transport and may become incorporated in a sediment forming either on land or at sea. Because of the high oxidation rates and rapidity of bacterial decay in the open air, land organisms, if preserved, are usually associated with water-laid deposits of some description. For continental areas this includes lakes and swamps and the parts of river courses where deposition is taking place. Also, animal tracks are more likely to become fossilized in areas where the animal has crossed a wet mud flat or similar feature, which has subsequently dried out.

The best chance of preservation occurs in the "mixed" environments of river deltas and estuaries. Here, where the velocity of the river water is checked, energy levels are usually low enough and the sediment load great enough for organisms to be smothered in sediment and fossilized. Such organisms may either be indigenous, or they might have been brought down from points higher up in the catchment area.

Marine environments

The majority of sedimentary rocks are of marine origin and the chances of preservation of fossils are far higher than on land. Within the marine areas of the modern world, several markedly different environments may be identified. These are shown in diagrammatic form in Fig. 2.5.

The diagram is not drawn to a consistent scale and it is important to note that sunlight can only penetrate, at most, the top 200 m or so of sea water. This is known as the <u>photic zone</u>. Contrast this with the average depth of the oceans of 3800 m and the depths of over 10,000 m in the deepest ocean trenches. The depths below 200 m are completely dark and are also considerably colder than most of the surface waters. Ultimately, organisms in any food chain depend upon plant life. Plants in turn need

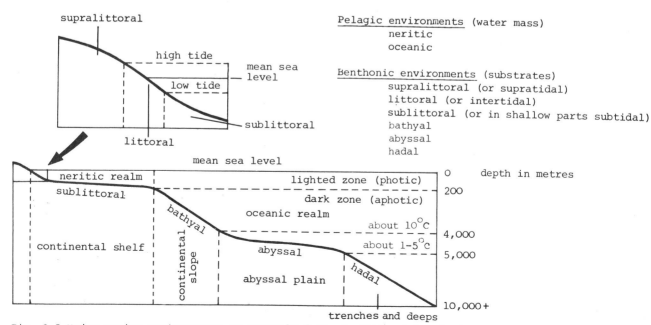

Fig. 2.5 Major marine environments as commonly defined. Although boundaries are not precise, note the general correspondence of sublittoral, bathyal, abyssal and hadal environments with continental shelf slope, abyssal plain and trench. The penetration of sunlight decreases with water depth, being virtually absent in waters deeper than 200 metres, where it is always dark.

abundant sunlight for photosynthesis, so the primary source of food in the oceans must originate in the photic zone. In the open ocean, most of this plant life consists of microscopic organisms known as phytoplankton. These form the food source of tiny animals known as zooplankton. Both forms of plankton are eaten by larger organisms in the photic zone, and so the sequence continues through the various food chains.

Naturally, dead organisms from the photic zone eventually descend towards the sea bed and form a possible food supply for further living creatures, either within the water mass or at the sea floor itself. Fig. 2.6 shows, in general terms, how the balance is achieved between the components of an eco-system. Fig. 2.7 puts flesh and blood to the theoretical "bones" and depicts a typical oceanic food cycle.

The marine fossil record, of course, contains only those creatures which reached the sea floor either before or after they died, so one sees a kind of "telescoped" view of a former vertical sequence of life forms. In working back to a reconstruction of the living scene from fossil assemblages, we usually try to distinguish those which actually lived in the water mass (pelagic), from dwellers on or in the sea bed (ben-thonic).

Pelagic organisms are clearly less restricted in their distribution than benthonic ones, since they can swim or float far above the vagaries of sea bed relief and sediment. Ocean currents, or their own swimming activity may carry them far and wide across the world's oceans.

It is only to be expected that the dark, cold

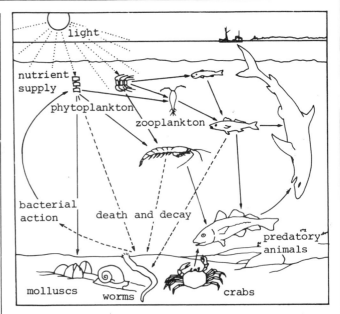

Fig. 2.7 The food cycle of the oceans.

and often anaerobic conditions of the deep-sea basins are not so conducive to life as the shallower water of the continental shelves and coastal areas (neritic and littoral zones, respectively). The lower depth limit of the continental shelves rough-ly corresponds to the base of the photic zone, so the shelves are, in general, well lit and oxygenated. Benthonic forms of life, both plant and animal, may therefore thrive and there is considerable diversity of ecological niches in such areas. These neritic and littoral zones are also underlain by continental crustal material, which is of lower density than that of the oceans. It is therefore more likely to be preserved in the geological record and not "lost" by sub-duction at the margins of tectonic plates.

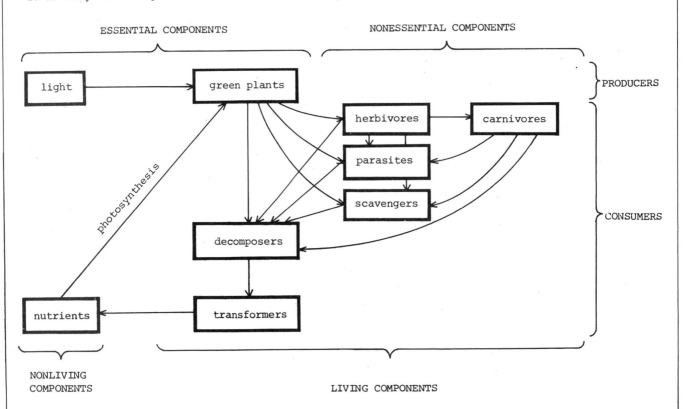

Fig. 2.6 Diagram of the flow of materials through a typical ecosystem.

3
Preservation of Organisms as Fossils

The Breakdown of Dead Organisms

One of the delights of geological field work is to discover well-preserved fossils. All of us have in mind some of the superb specimens which we have seen in museums, but it seldom falls to our lot to find similar ones. Neither are there many localities where one can find as great a diversity of species all together as may be seen on a modern beach. Why is there such a discrepancy between the living world and the fossil localities?

Some of the reasons are not difficult to visualise and they may be grouped as follows:

Biological breakdown

It does not take long for the soft parts of an animal to decay by bacterial action or by the work of scavengers. Very seldom do we find any trace of the soft body in shells picked up at the seaside, and to find evidence of soft parts preserved in fossils is rarer still. However, the geological record contains some celebrated exceptions. One example is the Ordovician trilobite *Triarthrus*, some specimens of which show the

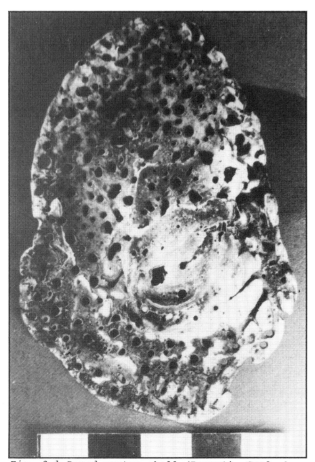

Fig. 3.1 *Bored oyster shell (Recent). Scale bar is 5 cm.*

remains of soft "appendages" preserved as a pyritized film on the bedding plane. These are usually interpreted as legs, with some form of gills attached to the upper ends.

The Solnhofen Limestone, from the Jurassic of Bavaria, has also yielded some superbly preserved fossils of a variety of animals, displaying many imprints of soft parts, which must have been impressed onto the substrate before decaying. It is generally believed that the sea bed conditions at that time were too foul for the survival of predators and other benthonic organisms which would have eaten the remains or disturbed them. The importance of the discovery in the Solnhofen Limestone of a birdlike creature, *Archaeopterix*, which even retains the imprint of feathers, is well known.

It is, however, not only the soft parts which are subject to biological decay. Fig. 3.1 is a photograph of a modern oyster shell which has been heavily perforated by boring organisms. The outcome is a specimen so weak that it will almost fall apart in the fingers, in contrast to the original tough shell. Several organisms are capable of such destructive activity, notably sponges, cirripedes, some algae and even other species of bivalves.

Mechanical breakdown

In a high energy environment such as a beach, the constant wave battering will obviously assist in the breakdown of dead organisms, especially when the water is "armed" with pebbles or sand. One of the outcomes of this is the production of a shell sand, but total destruction of shell and other material is not inevitable. Instead, some groups of organisms may be more susceptible to breakdown than others and the resulting assemblage of fossils may give a very biased record of what actually lived.

A fair imitation of the activity of waves may be made by first weighing a variety of modern shells and then shaking them with a handful of pebbles in a plastic bottle, such as a 2 litre orange squash container. The bottle is shaken violently for one minute and the contents tipped out and examined. Complete shells or fragments which are bigger than 5 mm are weighed again and plotted as a percentage of the original. These are replaced in the bottle and the process repeated many times until the bulk of the material is destroyed or the operator is exhausted! Clearly, the fossil remnants of such an assemblage would be very different from the original. The simple experiment uses only bivalve and gastropod shells, and seaweed, the weakest of which is strong compared to organisms such as echinoids. (See Fig. 3.2).

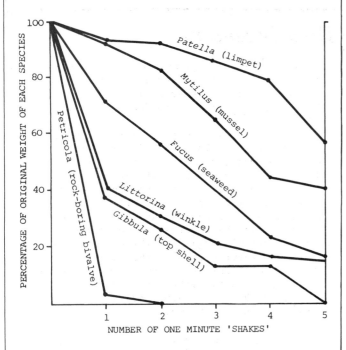

Fig. 3.2 Mechanical breakdown of beach material in a home-made 'shaker'.

Chemical breakdown

Biochemical processes have already been referred to. In addition, calcareous material is subject to solution, although this is most common in acid groundwaters after the organism has been fossilized. There may therefore be some visible imprint such as a brachiopod mould in the rock matrix, which in many cases is as valuable as the shell itself, and is still technically known as a fossil.

The limited nature of the fossil record

When we also take into account the recycling of fossiliferous sequences by erosion or by metamorphism we may count ourselves fortunate that we can find any fossils at all! Indeed, a somewhat tentative calculation has been made which suggests that the fossil record from the base of the Cambrian to the present day may contain as little as 0.01% of all the species which could have lived during that time! However unlikely this figure may be, we do well to remember that:

a) Most fossil assemblages are an incomplete record of the original living community, and reconstructions of the former environment must be made with great care.

b) The total fossil record is very incomplete and evolutionary lineages worked out from it are subject to frequent revision as new material comes to light.

Assemblages of Fossils

We have already noted the contrast between finding living organisms in a rock pool and empty wave-washed shells lying at the water's edge. The palaeoecologist is equally concerned to know whether the fossil assemblage he is studying represents organisms which were buried and fossilized more or less where they lived, or whether, instead, they were washed about and possibly transported far from their positions in life. The first type of assemblage is known as a life assemblage, the second as a death assemblage. In each case, only a fraction of the complete community of once-living organisms is likely to be preserved at all, but for both kinds of assemblage, the kind of environmental reconstructions which one can make will be quite different.

In many situations of course, mixed assemblages occur: to revert to our modern sea shore, the dead bivalve shells may be separated by only a few centimetres of sand from living ragworms or echinoids in their burrows beneath. Fig. 3.3 below summarises the ways in which life and death assemblages may be formed.

Observations of palaeoecological value may be made on the whole assemblage in situ in the field, or on the smaller scale in the laboratory after specimens have been "cleaned up". It is difficult to imitate the field situation

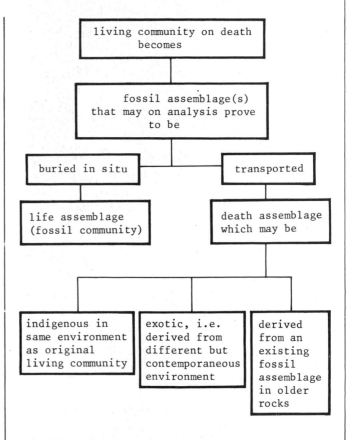

Fig. 3.3 Classification of fossil assemblages according to mode of origin.

in a book, but you should find it possible to deduce something about the history of the fossils or of the sedimentary matrix in each of the photographs (Figs. 3.4 to 3.11). For example:

Are there any close relationships between organisms fossilized together? If so, are they of a symbiotic nature, or did one animal take advantage of the dead body of another to use it as a micro-habitat?
Can you distinguish any relationship between the fossils and the sedimentary matrix? If so, can you determine the former energy levels, water depth or continuity of sedimentation?
Which specimens represent life assemblages and which are death assemblages?

In each case, brief descriptions are given beneath the photographs, with the original orientation where known. Suggested answers are also given at the end of this section, but do try the exercise for yourself first! If you wish to be more detailed, you could apply appropriate parts of the questionnaire on page 21 to the photographs. Each division on the scale represents 1 cm.

Note that in your own field investigations of fossiliferous exposures, you are likely to learn far more by examining the fossil assemblage as a whole, rather than trying to extract a "museum specimen" free of any matrix or associated fauna.

Fig. 3.4 *Bryozoa, brachiopods and a trilobite cephalon in a muddy limestone. Wenlock Limestone, Silurian, Much Wenlock, Shropshire. View of the top of the bedding plane.*

Fig. 3.5 *Lonsdaleia, a colonial coral. Carboniferous. Locality unknown. The upper part of the specimen is preserved in chert; the lower in crystalline limestone. Assume that this is the original orientation.*

Fig. 3.6 *Spondylus, a bivalve. Cretaceous. Warlingham, Surrey. Only the spine-bearing right valve is preserved. One spine is still intact.*

Fig. 3.7 *Orthid brachiopods, tentaculites and a trilobite cephalon (to left of scale). Ordovician. Marshbrook, Shropshire. Decalcified specimens in a yellowish sandstone. The brachiopods are disarticulated. View of top of bedding plane.*

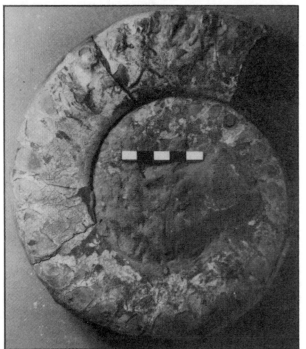

Fig. 3.8 Apoderoceras, a large ammonite. Lower
Jurassic, Scunthorpe. Note the worm tubes encrusting the specimen. The shale matrix includes
fragments of bivalve fossils of 'active' type.

Fig. 3.10 Bivalves of the <u>Mya</u> family in a
calcareous mudstone. Lower Jurassic, Blockley,
Oxfordshire. Both valves are preserved in each
case. The scale marks the base of the specimen
as found in the quarry face.

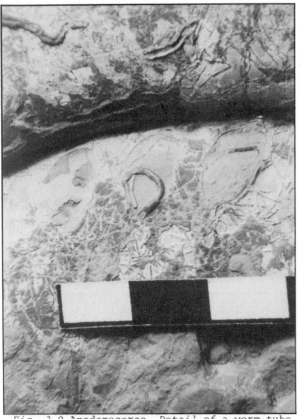

Fig. 3.9 Apoderoceras. Detail of a worm tube
apparently cemented to the matrix protruding
through a broken tubercle.

Fig. 3.11 Trace fossil. U burrow made by an
annelid worm or a crustacean. Middle Jurassic,
Ravenscar, Yorkshire. The matrix is a fine
sandstone. Assume that this is the original
orientation.

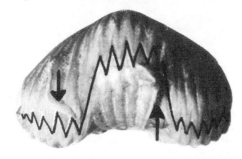

Fig. 3.12 Goniorhynchia, a rhynchonellid
brachiopod. Middle Jurassic, Dorset. The
zigzag opening line has been picked out in black.
Note the calcareous worm tubes on each valve
(arrowed). Width of specimen, 3 cm.

Suggested answers

Fig. 3.4. Wenlock Limestone, some fragmentation of the weaker specimens, such as the bryozoa has occurred here, but the brachiopods and the trilobite cephalon are mostly complete. This would suggest moderate energy levels. This is a death assemblage of shallow-water marine origin.

Fig. 3.5. *Lonsdaleia*. The chert preservation provides a rare opportunity to view the delicacy of the coral structure. The rest of the specimen is preserved in calcite. Assuming the orientation stated in the caption, the specimen must be the right way up and represents fossilization in the life position.

Fig. 3.6. *Spondylus*. The preservation of a delicate spine indicates that the shell cannot have moved far before being fossilized. In life this valve would have been the lower one, by which it was anchored to the soft substrate. However, the left valve is missing, so there must have been some water current activity on the sea bed.

Fig. 3.7. Ordovician assemblage. This is a marine death-assemblage, from the disarticulated state of the brachiopods and the trilobite. The mode of life of the tentaculites is unknown but they were probably benthonic, like the brachiopods and trilobite. The brachiopods are all 'convex-up', suggesting moderately high energy levels. The tentaculites seem to have a slight preferred orientation (up and down the page) suggesting sea-bed currents in that direction. The outline of the brachiopods being nearly circular, shows little effect of current orientation.

Fig. 3.8 and 3.9 *Apoderoceras* (a large ammonite). The ammonite and associated bivalves represent a death assemblage, since ammonites were nektonic and the bivalves benthonic. The interest of this specimen lies in the close-up of the worm tube. This appears to cross from the shell to the shale matrix where a tubercle has broken off. Such worms needed a hard substrate to live on but it is difficult to see how the matrix could have become lithified before the whole assemblage was fossilized. The following hypotheses might be put forward: (i) that the worm tube rests on a thin sliver of shell rather than on the matrix; (ii) that the ammonite was 'derived' by erosion from a slightly older horizon and was already a fossil before the time at which the worms lived; (iii) that the ammonite became fossilized beneath a cover of sediment. This sediment was later removed by erosion down to the level of the ammonite, which formed part of the sea floor long enough for the worms to secrete their shells onto it. No doubt you can think of other ideas which have just as much chance of being correct as any of ours!

Fig. 3.10. Burrowing bivalves. The elongated shells are typical of the bivalves which burrow into soft sediment. They are aligned parallel to each other in a near vertical position so it is quite likely that this part of the assemblage is a life assemblage. The rest of the block of mudstone contains a valve of an active bivalve, one of a mussel and a small ammonite (not visible in the photo) so the whole assemblage is quite a mixed one.

Fig. 3.11. U-burrow. The creature which once occupied the burrow lived in sand which was still soft, evidenced by the slight semi-circular marks, left between the branches of the tube as the animal deepened its burrow. This could have been in response to erosion of sediment from above. Modern U burrows are usually characteristic of very shallow water, close to or even above low tide mark. Like most trace fossils, the specimen was probably fossilized in situ.

Fig. 3.12. *Goniorhynchia*. Each worm tube seems to stop short at the actual line of opening of the brachiopod, suggesting that the worms were possibly enjoying the benefit of extra food supplies brought close by the brachiopod's powerful water-current mechanism. (This may however be purely fortuitous in this case, since the bed from which it comes is known to be a 'condensed' sequence, where dead brachiopods could very well have provided a local hard substrate for worms).

4
Reefs Ancient and Modern

Introduction

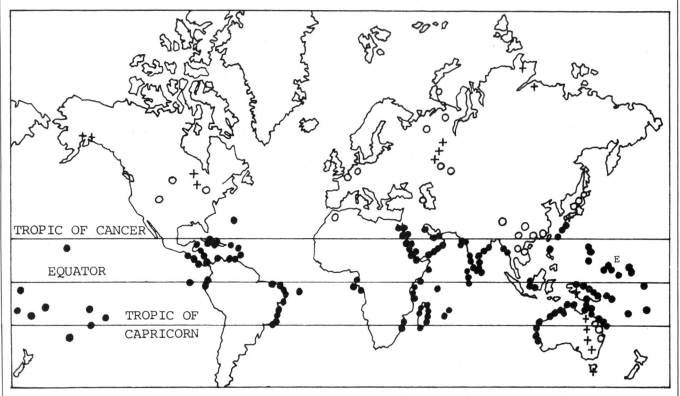

Fig. 4.1 The distribution of recent and fossil reefs. Black dots represent Recent coral reefs; circles, Lower Carboniferous reefs; crosses, Silurian reefs; 'E' is Eniwetok Atoll.

The very word "reef" seems to be an emotive one! To the layman it conjures up an image of sunlit tropical seas, dotted about with coral atolls, or of tourists examining the Great Barrier Reef of Austrialia through glass-bottomed boats. To the geologist, a "reef" is a characteristic complex of sedimentary rocks containing diverse fossil organisms, which many of his colleagues will argue was never a reef at all!

It is the purpose of this section to outline the main features of a modern reef and of a Palaeozoic one and to see how much they really do have in common.

Most modern reefs are built of coral, the conditions for the growth of which are well known. These include: water depth of less than 90 m, water temperature of between 18°C and 29°C - preferably near the warmer end of this range, "normal" salinity (i.e. about 36 parts of dissolved salts per thousand)

and clear water which is devoid of detrital sediment.

The main reason for such closely circum-scribed conditions is that most reef-building corals grow in close symbiotic relationship with algae. Algae, being plants, need the sunlight in order to photosynthesise their food.

The map (Fig. 4.1) shows the distribution of modern coral reefs; they are generally most abundant in the Pacific and Indian Oceans and elsewhere in the Tropics where warm currents impinge upon the east coasts of continents. The Pacific and Indian Oceans contain the most diverse range of corals of all types, over 700 species having been recorded there. Corals of non reef-building type may occur in a wide range of conditions: some thrive in deep water, others in more turbid waters and solitary corals are even known in waters as cool as those off the Norwegian coast.

Eniwetok Atoll – A Modern Reef

One of the best known reefs is Eniwetok Atoll in the Pacific Ocean. The living reef rests

on some 1250 m of limestone, composed largely of dead coral skeletons and debris and apparently formed by nearly continuous coral growth around a slowly subsiding volcanic island during most of the last 65 million years or so.

Fig. 4.2. is a general section through an atoll of this type. Detailed studies of the distribution of organisms across the reef reveal a fascinating pattern, similar in some respects to the seaweed survey in Section 2. Many types of organism are worthy of study, but the ones chosen here are the reef-building corals.

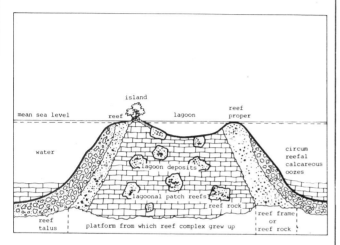

Fig. 4.2 A generalised section through a coral atoll.

The lateral distribution of the living coral across the windward reef at Eniwetok is shown in Fig. 4.3. Whilst some coral species seem to tolerate a wide range of current/energy levels, others can only survive in a very closely controlled environment. For example, *Turbinaria globularis* is limited to the quieter back-reef areas where it is sheltered by the bulk of the reef, whilst *Pocillopora danae* is only able to thrive where the surf is breaking.

Naturally, storm action and the activities of predators such as the "Crown-of-Thorns" starfish result in some of the coral and other skeletons being broken and ground down. Some of the resultant calcareous waste falls down the face of the reef towards the deeper water to form an underwater scree, or <u>fore-reef</u>. The slope is steep and much of the carbonate debris may accumulate at the maximum angle of rest, namely about 32°. If binding agents such as sea-grass, bryozoa or other encrusting organisms are present, the slope may be steeper.

On the sheltered side of the reef, or <u>back-reef</u>, the initial slopes are gentle and the carbonate debris comes to rest in a more nearly horizontal state, where it may be re-worked by sediment-eating or burrowing organisms.

Fig. 4.3 Cross-section of the windward reef of ▷ Eniwetok Atoll, showing current velocity gradient and species zonation.

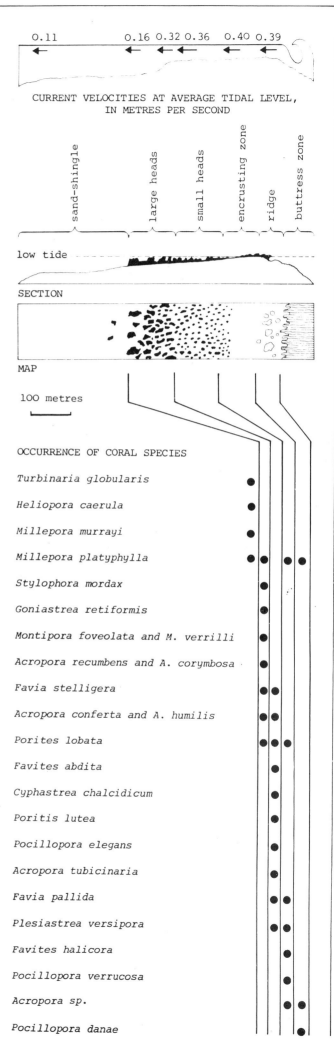

25

Fossil Reefs

The geological column, from the Palaeozoic to the Recent, contains many structures which are often interpreted as ancient reefs. The geographical distribution of these reefs is very widespread, and some are shown in Fig. 4.1. The most notable examples in this country are the Silurian "ball-stone" reefs of the Welsh Borders, reefs of several different types in the Lower Carboniferous of the Pennines, and the "patch reefs", or "thickets" of the Jurassic of Oxfordshire. We shall concentrate on a Lower Carboniferous example.

Perhaps the closest parallel in structure to the modern reefs is the so-called "apron-reef" of the Carboniferous, although the tectonic setting of the underlying rocks at the time of deposition was quite different. Like the atolls, an apron reef consisted of a wall constructed by reef-building organisms with a steep underwater slope on one side (the fore-reef) and a more gentle back reef on the other. A stylised section through such

a complex is shown at the base of Fig. 4.4. Most of these reefs developed along the edges of resistant blocks of older rocks, which underly the South Pennines and the Askrigg district of the North Pennines. In Lower Carboniferous times, these underwater blocks were apparently steadily subsiding as fast as carbonate material accumulated on top of them. Adjacent areas sank rather more quickly, forming miniature basins and the apron-reefs were produced along the junction between the two areas.

Although the structure may be broadly comparable to that of the atoll, the reef-building organisms are not necessarily the same at all. One major contrast is that the types of corals in existence in the Palaeozoic were of very different construction from those of today. They consisted of colonial tabulate corals and colonial and solitary rugose forms, all of which had become extinct by the late Permian/lowest Trias. The evolutionary links between these and their

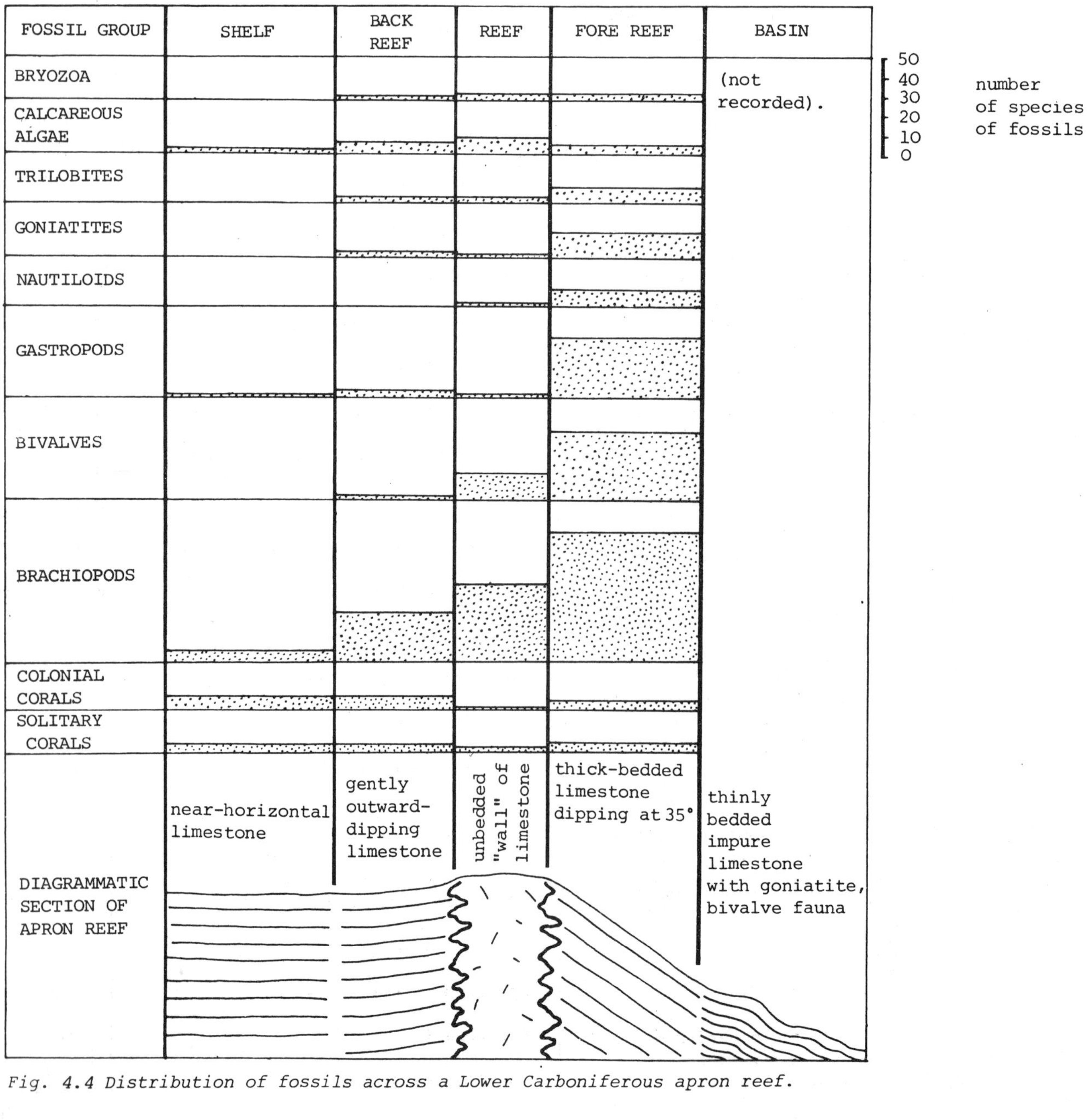

FOSSIL GROUP	SHELF	BACK REEF	REEF	FORE REEF	BASIN	
BRYOZOA					(not recorded).	
CALCAREOUS ALGAE						
TRILOBITES						
GONIATITES						
NAUTILOIDS						
GASTROPODS						
BIVALVES						
BRACHIOPODS						
COLONIAL CORALS						
SOLITARY CORALS						
DIAGRAMMATIC SECTION OF APRON REEF	near-horizontal limestone	gently outward-dipping limestone	unbedded "wall" of limestone	thick-bedded limestone dipping at 35°	thinly bedded impure limestone with goniatite, bivalve fauna	

50
40
30
20
10
0

number of species of fossils

Fig. 4.4 Distribution of fossils across a Lower Carboniferous apron reef.

Mesozoic to Recent successors are still not properly understood and it must not be assumed that the Palaeozoic corals required the same conditions.

Another problem is that in dealing with a fossil assemblage, one must expect that some species in the original community will not be represented. For example, on the modern sea floor, sea grass and soft-bodied algae play an important part in binding together loose sediment, but they would not be preserved as fossils.

In spite of these problems, attempts have been made to record the very rich and diverse fauna and flora of the Carboniferous reefs with a view to reconstructing their former state. The following example consists of an abstract of such work and if you follow it through carefully you should be able to work out your own reconstruction of a Lower Carboniferous reef.

Fig. 4.4 shows the results of painstaking collecting and research in North Derbyshire and North Staffordshire, where reefs of Lower Carboniferous age are well exposed. The data are shown in the form of histograms (bar graphs) of the number of species of fossils found in each section of the reef belt. (The basin facies limestone are not described, since these are obscured by younger rocks in the immediate vicinity of the reefs.) The graphs are thus expressions of the diversity of fauna and algae which were fossilized in the different zones, but they do not necessarily show the distribution of the living forms. Why not? Note that the diagram shows the variety of species, not the relative abundance of specimens.

It is tempting to compare such plots directly with the study of Eniwetok Atoll, but to do so would be premature. We must first try to reconstruct the living positions of the organisms and the original structure of the reef, since the former might have been disturbed before fossilization and the latter might have been affected by earth movements.

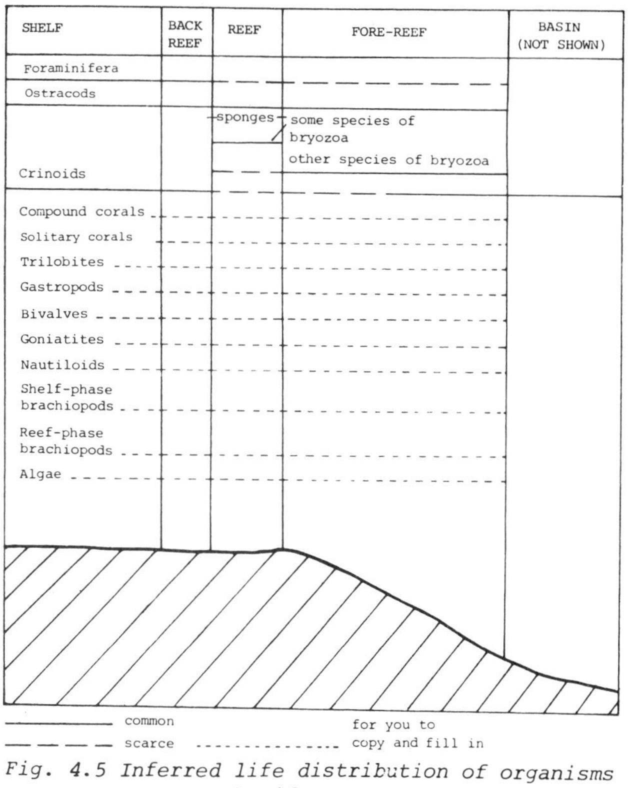

You should try this for yourself. Copy Fig. 4.5 and then try to complete the lines to show the inferred life distribution of the organisms across the reef belt. Use the symbols shown in the key. Some organisms have already been plotted, using data in the original published papers. You should find enough information to finish the diagram from Fig. 4.4 and the data below:

Data

The modes of life of the organisms listed in Fig. 4.5. are generally as follows:

1. Benthonic filter feeders: bryozoa, bivalves, brachiopods, colonial and solitary corals. Benthonic plants growing by photosynthesis: algae. Benthonic scavengers or sediment eaters: trilobites, most gastropods. Nektonic predators: nautiloids and goniatites (possibly swimming closer to the sea bed than modern nautiloids).

2. Colonial and solitary corals in the reef limestones are unworn and are mostly in their positions of growth. In the fore-reef, some solitary corals are in growth positions, but colonial ones are broken and water-worn.

3. The state of preservation of the trilobites and gastropods suggests that they mostly lived where they are now found.

4. Bivalves are uncommon in the shelf and back-reef facies. In the reef, well-preserved single shells of byssally fixed, burrowing and free-living forms are found. The biggest variety is found in the fore-reef, where many bivalves still have both shells present. A possible rock-borer is recorded from the reef and the fore-reef.

5. Brachiopods are abundant in the reef-complex and many of them seem restricted to particular habitats, like the corals at Eniwetok. Brachiopods characteristic of the shelf do occur in the reef, but only as single, drifted valves. Typical reef forms occur well preserved in the reef and in a broken state in the fore-reef. However, not all the brachiopod fossils of the fore-reef have fallen from the reef itself, since many are beautifully preserved in every detail. Recent work has demonstrated that the fore-reef slope provided several micro-habitats, some brachiopod species being found only at closely defined former water depths.

6. In North Derbyshire, some of the brachiopod shells in the fore-reef may be used as "fossil spirit levels". They were fossilized in their growth positions and part of each empty shell became filled with carbonate mud, the top surface of which would have been horizontal at the time. These mud bedding planes within the fossils now dip at an

SHELF	BACK REEF	REEF	FORE-REEF	BASIN (NOT SHOWN)
Foraminifera				
Ostracods				
Crinoids		sponges	some species of bryozoa	
			other species of bryozoa	
Compound corals				
Solitary corals				
Trilobites				
Gastropods				
Bivalves				
Goniatites				
Nautiloids				
Shelf-phase brachiopods				
Reef-phase brachiopods				
Algae				

——— common
– – – scarce
.... for you to copy and fill in

Fig. 4.5 Inferred life distribution of organisms across a Lower Carboniferous apron reef.

average of about 7°, whilst the dip of the bedding planes of the limestone itself is about 35° in the same direction. Clearly, post-Carboniferous earth movements have affected the rocks by only 7°; the other 28° dip represents the original slope of the fore-reef surface. (See Fig. 4.6).

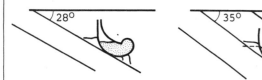

Brachiopod, fossilized in growth position. Mud partly fills shell.   Limestone sequence and brachiopod tilted 7° by earth movements.

Fig. 4.6 The formation of a fossil 'spirit level'.

7. Where calcareous algae are recorded from the reef, they are in their positions of growth; elsewhere, they are not.

8. Crinoid stems are important constituents

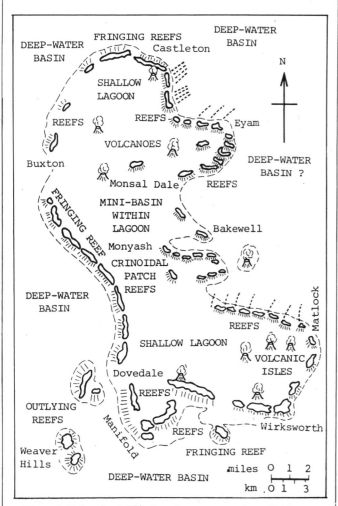

Fig. 4.7 A simplified palaeogeography of the Peak District in Lower Carboniferous times.

of the assemblage on the shelf and in the lower slopes of the fore-reef. They are not common in the reef itself, nor on the upper slopes of the fore-reef.

When you have completed a copy of Fig. 4.5, compare your findings with Fig A.1 (Appendix). This is an artist's reconstruction of the possible appearance of the reef in Lower Carboniferous times, although of course the organisms and the reef are not drawn to the same scale. Organisms are shown in their life positions on the sea bed and within the water mass and as they appear in fossil form within the sediment.

Your completed version of Fig. 4.5 and the block diagram (Fig. A.1) will enable some comparison to be made with the modern Eniwetok reefs. Try to make this comparison yourself, noting particularly which organisms are the main reef builders in each case.

Fig. 4.7 is an attempt to incorporate all the data of the reef belts of the Carboniferous Limestone block of the South Pennines into one composite diagram. This is acknowledged to be an over-simplification, since the reefs were not all strictly contemporaneous, but it does emphasise the consistent contrast between the stable <u>platform</u> occupied by the lagoon and the more rapidly subsiding <u>basins</u> beyond the reef belt. It is similar in shape and size to some modern atolls. Unlike Eniwetok the underlying platform is not a former volcano, but consists of a larger block of older rocks, probably deposited in Ordovician times or before.

We are now in a better position to reconsider the latitude of Britain during the Lower Carboniferous. The Carboniferous reefs are largely built by <u>algae</u>, not coral, so the water must presumably have been shallow and well-lit. Some of the algae seem to have thrived in very high energy levels, like some modern corals at Eniwetok.

The low proportion of detrital sediment in the limestone demonstrates that the sea water was clear. Life in the ancient reefs was diverse and abundant: some of it consisted of coral, albeit of a different type from today. These factors all lead us to suppose either that Britain was in more tropical latitudes in the Carboniferous, or that the world climate was warmer. There was, however, ice in the polar regions during the Carboniferous, so perhaps a lower latitude for Britain is a better explanation. It is no suprise, therefore, to discover from palaeomagnetic measurements that the latitude of central Britain at the time was within a few degrees of the Equator.

5

The Frodingham Ironstone

Introduction

Within the Jurassic and Cretaceous of Western Europe a variety of bedded ironstones occur. Although obviously of sedimentary origin, the exact ways in which they were formed are difficult to determine. Can we, perhaps, use the fossils which occur within the ironstone to help unravel the mystery? A good example of such a rock is the Frodingham Ironstone, which is found near Scunthorpe, in South Humberside. Similar deposits also occur near Corby in Northamptonshire and elsewhere in the Midlands but this account will concentrate on the Frodingham Ironstone which has been intensively studied (by Professor A. Hallam) from a palaeoecological point of view.

The ironstone is extracted from several enormous quarries by British Steel, who currently take about one million tonnes per year to blend with imported ores for the Scunthorpe blast furnaces. The workable deposit consists of a layer of ironstone approximately 9 m thick, extending for about 10 km in a north-south direction and, at depth, for a distance of some 30 km to the east. It is economically accessible where it forms a dip slope to the north and east of Scunthorpe (Fig. 5.1). To date it has proved worth working below a thickness of up to 60 m of waste rock, this overburden being removed by huge walking draglines, power shovels and dump trucks (Fig. 5.2).

The ironstone is of Lower Liassic age (Lower Jurassic) and most of the overburden consists of dark Liassic shales.

Fig. 5.1 Block diagram of the Frodingham Ironstone field. The main quarries worked in recent years are labelled in the diagram.

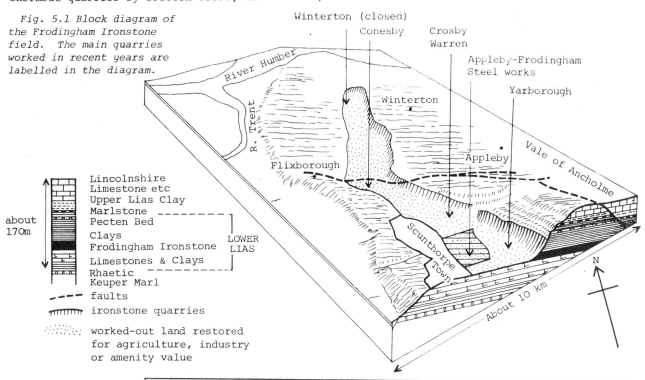

about 170m

Lincolnshire Limestone etc
Upper Lias Clay
Marlstone
Pecten Bed
Clays
Frodingham Ironstone
Limestones & Clays
Rhaetic
Keuper Marl

LOWER LIAS

- - - - faults

ironstone quarries

worked-out land restored for agriculture, industry or amenity value

Fig. 5.2 An ironstone quarry near Scunthorpe. The 'walking dragline' is sited on the top surface of the Frodingham Ironstone. Its boom is 92 m long.

Description of the Ironstone

The rock "as mined" has an average iron content of 22%. This is regarded by the steel-makers as a very low figure, but fortunately the ore contains a high proportion of calcium carbonate. It is blended with imported ores of a more siliceous nature and the high lime content of the Frodingham material makes the mixture self-fluxing, so no further lime needs to be added.

The iron minerals themselves comprise three main types:
<u>Siderite</u> (iron carbonate)
<u>Limonite</u> (a hydrated iron oxide)
<u>Chamosite</u> (a hydrated iron silicate)
The relationships between the three are quite complex and a variety of ironstones may be found:

TABLE 2 VARIETIES OF IRONSTONE IN THE FRODINGHAM IRONSTONE

Matrix	Particles
siderite/chamosite	limonite-chamosite ooliths
calcite - - - - -	limonite ooliths
siderite - - - - -	chamosite mud

Where limonite is abundant the ironstone is ginger-brown in colour; if siderite and chamosite predominate, it is grey-green.

The Problem of the Origin of the Ironstone Minerals

It is this very diversity of iron minerals which makes the deposit such an interesting one. In the laboratory, we can show that limonite normally forms under oxidizing conditions whilst chamosite and siderite form in quite markedly reducing conditions (i.e. absence of air). How can we have the three so inextricably combined? Unfortunately, there do not seem to be any similar ironstones forming today so we cannot directly apply the principle of Uniformitarianism to help solve the problem. Instead, we must use all the clues available in the rock itself, both sedimentological features and details of the fossils it contains.

Sedimentary Characteristics

The abundance of ooliths presumably indicates rather high energy shallow-water conditions, by analogy with carbonate ooliths forming on the Bahamas Bank today. However, the siderite-chamosite mudstones are very fine-grained deposits which would normally be identified with a low energy environment where chemical conditions were reducing. Although there are discrete layers of limonite oolites and siderite-chamosite mudstones, some of the latter also occur as separated clasts, apparently eroded from some already hardened layer and caught up in the next phase of oolite deposition.

Faunal Characteristics

Apart from the siderite-chamosite mudstones which are barren, the Frodingham Ironstone is rich in fossils; not only are there many specimens, but there is great diversity of species too. The list of recorded invertebrate species includes:

Ammonites	11	
Nautiloids	1	
Belemnites	1	
Bivalves	20	- representing most of the common modes of life.
Gastropods	2	(at least)
Brachiopods	3	(at least)
Crinoids	1	(at least)
Starfish & brittle star	2	
Foraminifera	1	
Trace fossils and sponges	several	

In addition, fossil wood and the traces of boring algae occur.
This list is similar to that which might be recorded at any "normal" Liassic shale locality, such as Robin Hood's Bay, on the Yorkshire coast, which would suggest that for most of the time in the Scunthorpe area the sea-water conditions were far from toxic! Indeed, some of the Frodingham specimens are larger than usual, ammonites up to 30 cm across being quite common and some of the burrowing and fixed bivalves reaching 20 cm or so. The ammonites do not give a true reflection of the sea bed conditions, since they lived in higher levels of the sea water, although one wonders how such big specimens came to be preserved intact in a deposit of normally high energy origin like an oolite. The specimen in Fig. 3.8 is from the "normal" Lias clays above the Frodingham Ironstone. (See section 3.)

The list in itself is interesting, but we are likely to obtain more information about the sediment as a whole if we examine the relationship between the benthonic organisms themselves and between the fossils and their sedimentary matrix. It is helpful, here, to

Fig 5.3 The top surface of a bedding plane in the Frodingham Ironstone.

use some sort of a questionnaire, to avoid missing important information.

Try using this simplified one on the slab of ironstone shown in the photograph Fig. 5.3.

Questionnaire for use with figure 5.3

About how many species are present?
What is the relative abundance of different groups?
Are any delicate structures preserved?
Are the specimens worn or broken? Some more than others?
Are the valves of bivalves separated?
Is there any preferential orientation to the specimens?

Conclusions Try to draw your own conclusions before you read on.

The usual interpretation of a slab such as this is that a "death" assemblage is represented and that moderately high energy levels operated on the shells to separate them and wash them uniformly into the stable "convex-up" position.

The more detailed questionnaire which follows is based on a published scheme (Ager 1976). You could provide yourself with a blank version of it to apply to a fossiliferous locality known to you. In this case, the table has been completed for the Frodingham Ironstone. Study it carefully in conjunction with the notes which follow and see if you can use the palaeoecological observations to help interpret the environment of deposition.

Additional observations

1. The deposit appears to be "condensed", i.e. it is thinner than the usual rock successions formed elsewhere during the same time span.

2. Many shells demonstrate the following relationship between the shell and the sediment (Fig. 5.4). This is taken to show

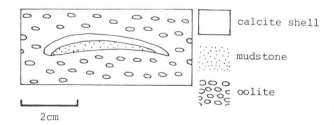

Fig. 5.4 Relationships between shells and matrix in the Frodingham Ironstone.

that the organism lived during a phase of mud deposition (as is the case with most of the Lias elsewhere). Later, the mud was winnowed away by current action and ooliths were deposited around the shell. However, underneath it, in the protected environment of the shell concavity, some of the mud was preserved.

3. Modern boring algae from the Black Sea mostly live in water depths of 20 to 25 m and never more than 40 m.

Palaeoecological Questionnaire

Use this to help you to study a fossiliferous horizon. Extra information should be shown by sketches, photos or sketch maps.

LOCATION SCUNTHORPE GRID REF.	GEOLOGICAL UNIT FRODINGHAM IRONSTONE (LOWER LIAS)	DATE LITHOLOGY CALCAREOUS IRONSTONE
1. Fossils in pockets, lenses, bands?	Mostly in uniform layers. Some small pockets of brachiopods.	
2. Even distribution through rock unit?	Abundant in limonitic rock: absent from chamosite - siderite mudstone.	
3. More abundant at any level in the unit?	Some fossils form beds, e.g. "Cardinia Bed".	
4. Same fossils as in same unit elsewhere?	Yes, virtually identical except for presence of many large fossils at Scunthorpe.	
5. Any species unusual, but common elsewhere?	One common bivalve species missing here.	
6. About how many species present?	See text	
7. Relative abundance of different groups?	See text	
8. Any obvious close associations?	Some small bivalves cemented to large Gryphaea shells.	
9. Any obvious derived fossils?	No	
10. Any encrusted or bored (fossils!)?	Borings by cirripedes, sponges and algae common in many shells.	
11. Any other fossils attached to each other?	See 8	
12. All fossils preserved in same way?	No. See note below	
13. Any delicate structures preserved?	Yes. e.g. echinoid spines	
14. Worn or broken? Some more than others?	Smaller shells at some horizons are broken Others not.	
15. Valves of bivalves separated?	Nearly always separated, except for deep burrowers.	
16. If still joined: tightly, part open, gaping?	Tightly	
17. Crinoid stems as isolated ossicles, or long?	Isolated	
18. Fossils in nodules preserved same as others?	Yes	
19. Any disturbance of sediment by organisms?	Many trace fossils, both "vertical feeders" or "horizontal feeders".	
20. Any in growth positions? If so, % of each species	Most of the Mya group (deep burrowers) are undisturbed in burrows.	
21. Any preferential orientation?	Disarticulated bivalves are "convex - up." Gryphaea "concave - up" (stable death position for Gryphaea)	
22. Borings, burrows, tracks, trails in sediment?	See 19.	

FURTHER NOTES OR PRELIMINARY CONCLUSIONS

12. Ammonites   - original aragonite altered to calcite

Bivalves etc   - many preserved in calcite

- some replaced by chamosite or siderite

- Mya group preserved as casts only.

Interpretation

No doubt, as you have been reading, you will have been working out possible explanations, but you might also wish to know what Professor Hallam decided!

He came to the conclusion that the Frodingham Ironstone was deposited on a large off-shore shoal. Many lines of evidence point to the water being shallow; it is unlikely, however, to have been a shoreline, since detrital minerals like quartz are uncommon. Energy levels must have varied considerably; at times they were low enough for mud to settle, at other times the sea bed was sufficiently lively for ooliths to form and for shells to be disarticulated and washed into their stable death positions. However, erosion was not too great, because burrowing bivalves and trace fossils are largely undisturbed.

If energy levels could fluctuate between low and high, perhaps oxidation-reduction conditions could have oscillated too. Professor Hallam seems to prefer this hypothesis to the suggestion that the variety in iron minerals could have been formed entirely by diagenesis, i.e. later chemical changes in the newly deposited sediment. We are still, however, left with the problem of how so many organisms could have flourished in waters which must have been abnormally rich in iron salts.

Fig. 5.5 is a reconstruction of the possible appearance of the sea bed during the deposition of the Frodingham Ironstone. The lower "ledge" of the block diagram represents the appearance of a quarry face in the Ironstone today.

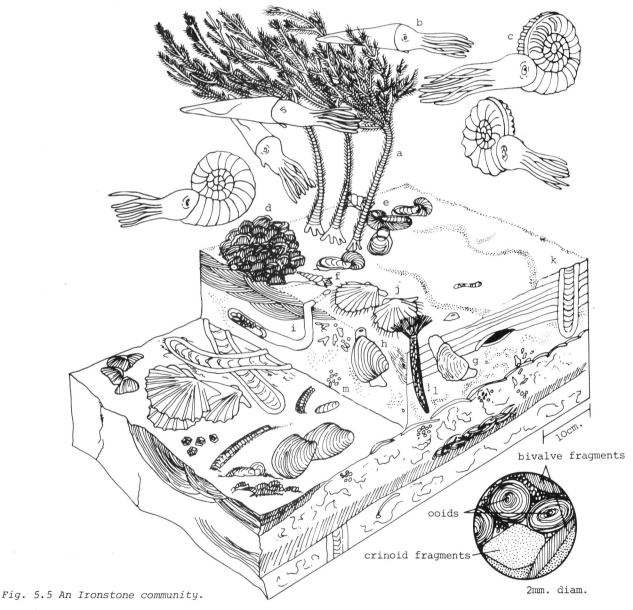

Fig. 5.5 An Ironstone community.

a Pentacrinus (Echinodermata: Crinozoa)
b belemnite (Mollusca: Cephalopoda: Coleoidea)
c Asteroceras (Mollusca: Cephalopoda: Ammonoidea)
d rhynchonellids (Brachiopoda: Articulata: Rhynchonellida)
e Gryphaea (Mollusca: Bivalvia: Pterioida-oyster)

f Procerithium (Mollusca: Gastropoda: Mesogastropoda)
g Pholadomya (Mollusca: Bivalvia: Anomalodesmata)
h Cardinia (Mollusca: Bivalvia: Veneroida)
i Rhizocorallium (trace-fossil - crustacean)
j Pseudopecten (Mollusca: Bivalvia: Pterioida - pectinid)
k Diplocraterion (trace-fossil - annelid or crustacean)
l terebellid (Annelida)
m Chondrites (trace-fossil - annelid)

6
Trace Fossils

Introduction

In previous sections of this book we have looked at how fossils can be used to reconstruct the conditions that existed in the past. Indeed, it is often possible to be very precise about the past environments using fossils as indicators and by comparing them with modern environments.

But what happens when the rock does not contain any remains of plants or animals? This does not necessarily mean that no life existed in the area when the sediment was being deposited; it could indicate that the plant or animal did not have suitable parts to be preserved, i.e. it could have been soft bodied, or the conditions under which deposition took place were not conducive to preservation. (Some of the reasons for non-preservation of organisms have already been discussed in Section 3 of this Unit.)

Fortunately, in these areas without animal and plant remains (body fossils) signs of life are not always completely lacking. Tracks, trails and burrows of the animals may be preserved and these <u>trace fossils</u> can give many valuable clues about the nature and habits of the creatures that made them. One of the features of trace fossils that makes them particularly useful in palaeoecological studies is that they are generally found in the sediment in which they were formed as they are rarely transported. That is, they are <u>autochthonous</u>. The same cannot be said of body fossils where it is quite possible for a specimen to be moved after death and transported into an environment in which it never lived, thus giving a misleading impression to the unsuspecting palaeontologist.

Trace fossils can be used as "way up" indicators as well as environmental indicators. A footprint leaves an impression on the surface and so a depression will indicate a top surface. On the other hand if the print has been filled in and is seen in relief this will indicate the undersurface of the bed. Many burrows can be used in much the same way.

The study of trace fossils is a branch of palaeontology known as <u>Ichnology</u> and just like other fossils, trace fossils are given generic and specific names. For example, *Diplocraterion luniforme* are vertical or U burrows found in Triassic sediments. There is a great variety of tracks, burrows and impressions throughout the stratigraphic column and it is often very difficult to decide what type of creature made them and what it was doing at the time.

In some cases this is quite an easy matter if the animal that made a burrow, such as a bivalve, is found in the burrow, or if at the end of a trail the animal that made it is found preserved. But it is rarely so easy to decide. Look at Fig. 6.1 which shows some

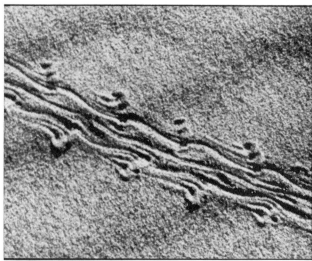

Fig. 6.1 A modern track. Can you decide what could have made these tracks across a modern sand dune (at Ravenglass) and in which direction it was travelling? (x 2/5)

modern prints on a sand dune in the Lake District. Could you work out what made the prints or even in which direction it was travelling? Turn over the page and look at Fig. 6.2 and see if you were close to the correct answer. The study of trace fossils presents even more problems than this modern example, for the animals that made the tracks may be extinct. Also, the trails of many animals such as bivalves, gastropods and worms are very similar and it may not be possible to decide which invertebrate group made them.

Clues to the origin of some trace fossils are given when their stratigraphic range has been found to match that of a particular fossil group. The origin of *Cruziana* (Fig. 6.3)

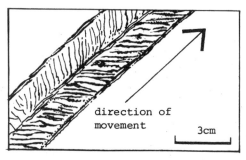

direction of movement
3cm

Fig. 6.3 A furrowing trail made by a trilobite.

after if was first described from South America in 1842 was a source of constant debate. Specimens were subsequently found in Palaeozoic rocks in many parts of the world. They were variously described as plants or the result of annelid or arthropod activity. However, the time range of *Cruziana* closely matches that of the trilobites and few people now would not agree that they were indeed the tracks of trilobites.

Cruziana have not been attributed to trilobites on their coincident stratigraphic ranges alone. Impressions of spines as well as parts of head and tail shields have been found with them which confirm the interpretation. You may be surprised at the direction of movement indicated on the diagram. The marks were made as the legs of the trilobite stretched forward and then were pulled back to its body as it dragged itself forward. This also produced a ridge of sediment in the centre of the specimen.

Fig. 6.4 *Tracks made across a bed of sandstone by a five-toed vertebrate animal; the specimen is a natural cast (on the underside of a bed) of original top surface imprints. The trace made by the tail dragging along the ground is clearly visible between the footprints. Direction of movement indicated by the arrow. Triassic, Alveley, Shropshire. (x 1/5)*

Some of the most spectacular trace fossils are the large reptilian foot prints that are found in Mesozoic strata. Fig. 6.4 shows the tracks of a vertebrate from the Triassic. Its size can be judged from the depth of the prints and from the distance between them. Whether it walked upright or on all fours can also be deduced. Some tracks show the drag marks of a tail and the size of any claws or webbing of the feet can also be seen. From these scraps of evidence, whole animals have been reconstructed even though no bones have been found with the prints.

Fig. 6.5 Cheirotherium. *(x 3/10)*

Cheirotherium (Fig. 6.5) is one such example. The hind and fore limbs differ, impressions of the fore limbs being smaller and not as deep as the hind limbs. It is only at a few localities that drag marks of the tail are found, so reconstructions of *Cheirotherium* show a medium sized dinosaur, similar to *Ornithosuchus*, which could move either on all four limbs or in an upright position and which generally carried the tail clear of the ground although it must have been quite long as drag marks are sometimes seen.

By anology with modern examples, certain types of trace fossil are likely to be confined to certain environments. Obviously vertebrate footprints will not be made in deep water!

Triassic Trace Fossils

One recent study that has used trace fossils to help establish a palaeoenvironment has been carried out in North Cheshire on beds of Upper Triassic age known as the Waterstones. Trace fossils were particularly useful in this case as apart from a few plant remains the succession lacks other fossils.

The sequence is a complex one of sandstones, siltstones, shales and mudstones showing a

variety of colours and which also show a great deal of lateral and vertical variation. In the past, the succession has been described as intertidal, estuarine or an inland, possibly saline, lake or sea. Deltaic, fluvial or freshwater influences have also been noted.

The lithology of the rocks and their sedimentary structures have had a great part to play in determining the environment. Structures such as mudcracks, ripple marks and pseudomorphs of halite, as well as the red colouration of many of the sediments in the sequence, clearly indicate some shallow water environments that periodically dried up.

The trace fossils found in the succession help us to obtain more precise ideas of the environment and the changes that occurred within it with time. These include *Diplocraterion* (see Fig. 3.11) and *Arenicolites* which are U shaped and vertical burrows. The *Arenicolites* are slightly larger, being up to 12 mm across. These burrows are interpreted as being made by invertebrate filter or suspension feeders, possibly an arthropod, or a worm in the case of *Diplocraterion,* and an annelid for *Arenicolites*. After studies of these trace fossils in a great number of different areas and by comparison with modern examples, they are taken to indicate intertidal conditions.

Other parts of the succession contain *Scoyenia* burrows which are a shallow, horizontal branching form. They are found in association with *Cheirotherium* prints which points to "non-marine" or marginal aquatic conditions, probably at the top of the tidal range or overlapping to freshwater or hypersaline conditions.

In the lower parts of the succession are branching burrows of *Thalassinoides*. Similar burrows of Jurassic age contain remains of a crustacean and likewise these burrows are thought to have been formed by a similar filter or suspension feeding crustacean. They probably indicate slightly deeper water than the other trace fossils described, possibly even being subtidal. Some of these burrows show signs of erosion and there are dessication cracks in the mudstone, (Fig. 6.6),

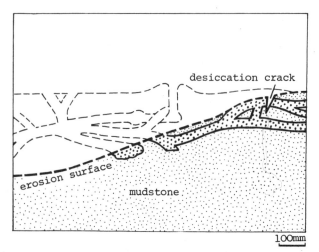

Fig. 6.6 Modes of preservation of burrow of Thalassinoides. *Mudstone and burrow system eroded and replaced by cross-bedded sandstone.*

so periods of sedimentation must have been followed by exposure causing some erosion of the beds containing the trace fossils.

Parts of the succession are lacking in trace

fossils but this is open to two interpretations. They could represent inhospitable environments, possibly because of high brine content or prolonged exposure, or it could be that they are not present because of the lack of a suitable casting medium.

The Waterstones succession is complex and variable, and when reconstructing the conditions that existed in the past, all the features of lithology and sedimentary structures, as well as the trace fossils have to be used. However, this is a case in which trace fossils have played a crucial role. All these factors taken together suggest that it was basically an intertidal area with subenvironments of exposed mudflats and sand bars in tidal channels.

Fig. 6.2 Natterjack toad making a track in a modern sand dune. How close were you? Perhaps you can now appreciate some of the problems which face palaeontologists trying to deduce what made tracks millions of years ago.

7
Microfossils

Introduction

Microfossils are an increasingly important branch of palaeontology. Micropalaeontology is the study of minute organisms which have to be extracted from sediment and examined with the aid of a microscope. These microfossils are every bit as diverse as the better known groups of macrofossils. They can be used in similar ways for dating and correlating the rocks.

However, the size of microfossils gives them advantages over the larger types in certain areas. One area which is very important today is in drilling operations of the sort carried out in oil exploration. Drilling breaks up the rocks into small chippings which are brought to the surface in the drilling mud and these chippings are sampled at regular intervals by the geologist. It is important for him to know what stratigraphic level the drilling has reached and he will need fossil evidence to date the rocks. It is extremely unlikely that he will recover a complete or large enough fragment of a macrofossil which can be used for dating, but in the chippings there could be hundreds of complete microfossils that can be used for this purpose.

As well as being useful for stratigraphic dating, certain of the microfossil groups, notably foraminifera, are valuable in indicating the environment of deposition of the rocks.

Foraminifera

Foraminifera are single celled animals that have existed from the Ordovician to the present day. They range in size from as little as 0.02 mm up to the giant foraminifera which can be 110 mm. Their composition also shows considerable variation. The shells or tests are either calcareous, or they are composed of sand or shell particles cemented together by an organic matrix. These are known as agglutinated foraminifera. There may be only one chamber or there may be a number of communicating chambers arranged in a spiral or a linear form. (Fig. 7.1).

Most of the foraminifera have been benthonic, but planktonic forms have occurred since the Jurassic and were important in the Tertiary. One calcareous, planktonic foraminifera, *Globigerina*, has made an important contribution to deep sea sediments since the Tertiary, being sufficiently abundant in certain places for the sediment to be given the name Globigerina ooze.

Coiling directions in foraminifera

In a study of deep sea sediments in the Atlantic Ocean, foraminifera preserved in cores from various localities were used by Ericson, Ewing and Wollin (1963) to try to establish the Pliocene-Pleistocene boundary. It was postulated that the differences noted in the planktonic foraminifera were caused by an abrupt climatic change that marked the onset of the first Pleistocene glaciation.

Previously it had been thought that the climatic transition between the Pliocene and the Pleistocene had been a gradual one, but the zone in which the marked changes occur is no more than about 10-15 cm thick which, at the rate of sedimentation taking place, must represent a time interval of about 5,000 years. In fact, the time interval may have been even shorter as some vertical mixing of the sediment by benthonic organisms may have happened.

These palaeontological changes include the appearance of *Globorotalia truncatulinoides* in abundance above the boundary and the extinction of *Globigerinoides succulifera fistulosa* at the boundary. But perhaps the

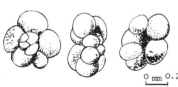

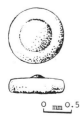

Globigerina
Calcareous test; planktonic
Palaeocene - Recent

Bolivina
Calcareous test, planktonic,
Upper Cretaceous - Recent

Saccamina
Agglutinated test; benthonic.
Silurian - Recent

Nummulites
Calcareous test; benthonic.
Palaeocene - Recent

Fig. 7.1 Foraminifera. Showing the variety of form, composition and mode of life.

most useful and spectacular change is that in the coiling direction of the trochospiral planktonic foraminifera *Globorotalia menardii* from 95% of the specimens being dextral (right coiling) below the boundary to 95% sinistral (left coiling) above it (Fig. 7.2).

This change in coiling direction does not appear to have any adaptive significance, but is the response of some genes to changes in temperature. Such a change in coiling direction has also been noted in other specimens of foraminifera such as *Globigerina pachyderma* but it cannot always be shown to have such a clear relationship with temperature change and perhaps it may in these cases be related to some other environmental considerations.

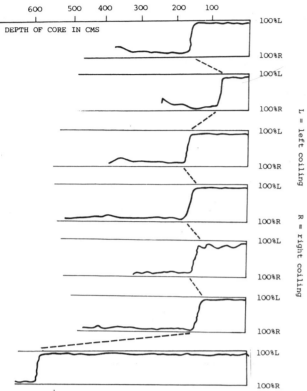

Fig. 7.2 *The Pliocene-Pleistocene boundary as indicated by changes in coiling direction of* Globorotalia menardii *from seven cores in the Atlantic Ocean.*

Benthonic foraminifera

Another example of the sensitivity of foraminifera to changing conditions is provided by benthonic foraminifera. In the Java Sea a strong zonation has been recognised over an area where the depth only varies from 30-50 metres and the temperature varies by as little as $\frac{1}{2}$°C. This zonation is controlled entirely by the nature of the substrate which does vary.

Six types of areas are delimited on the basis of variations in the substrate:

1. Lagoons - Few species. Small forms which feed on seaweed and dead corals.
2. Coral Reefs - Greater variety. Larger forms live among active corals and seaweed.
3. Sandy Reef Apron - Completely different population including giant foraminifera.
4. Mud Flat near Reef - Giant foraminifera and spinose forms, some with flattened tests.
5. Mud Flats far from land - Small foraminifera only, showing greater uniformity.

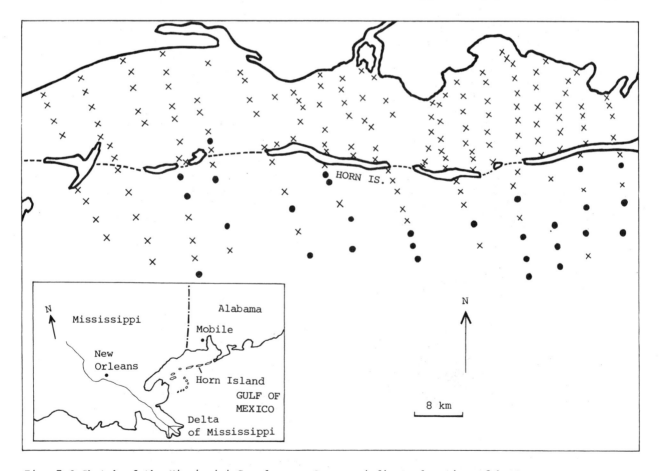

Fig. 7.3 *Sketch of the Mississipi Sound area. Crosses indicate location of bottom samples which failed to contain tests of planktonic foraminifera. Dots indicate location of bottom samples which contained tests of planktonic foraminifera.*

6. High energy coarse sandy bottoms - A few
   small, short-lived foraminifera.

Foraminifera in the Gulf of Mexico

The two examples above show how sensitive
foraminifera can be to a variety of changing
conditions; but can this knowledge have any
useful applications? A recent study of the
distribution of modern foraminifera in the
Gulf of Mexico was funded by oil companies.
They wanted to apply the research to their
quest for oil.

A series of sample transects was taken from
the shore going out into deeper water beyond
a chain of off-shore islands (Fig. 7.3). The
benthonic foraminifera are more concentrated
in the lagoons inshore from the islands which
stop the sea invading, whilst the planktonic
foraminifera are not found in these areas.
Beyond the islands in the open water the
abundance of the planktonic foraminifera
starts to increase and they reach up to 70%
of the total population. It is possible to
be very precise about the distribution of
certain forms, for example *Ammonia becarii* is
found in shallow brackish waters where

streams enter the lagoon and *Uvigerina* is a
form that comes in cooler, deeper water down
the slope. But even at a very simple level it
is possible, just by using the appearance and
build up of the planktonic foraminifera, to
note the changes that take place around the
islands.

Obviously modern sand bars and barrier islands
can be spotted from their shape but the oil
companies wanted to identify them in the
stratigraphic record, as these types of
environment can give rise to oil traps.
Because the same relationship between the
planktonic and benthonic foraminifera has
probably held since the Upper Cretaceous, the
work was applied to Tertiary sediments in
Louisiana and Mississippi, U.S.A. In these
areas a series of transgressions and
regressions could be recognised based upon
the assemblages of foraminifera and this
enabled the oil companies to correlate former
environments within the area to find the most
likely sites in which to sink their wells.
Foraminifera are not the only microfossils
useful in palaeoecological studies. Pollen
and spores also tell us a lot about former
environments.

Pollen

In the recent past, fluctuating climates in
the Pleistocene have caused a change in the
vegetation. A record of these changes is
preserved in the sediments by their pollen.
Fig. 7.4 shows clearly the relative abundance of the different species of pollen
extracted from an interglacial lake deposit
of Pleistocene age.

The late glacial stages of the Lowestoftian
pass into the early stages of the Hoxnian
interglacial which show a pretemperate
climate, birch being the dominant tree. As
the climate warms up, species such as oak,
alder and hazel make their appearance. The
pattern is reversed as the climate cools and
the early Gippingian glacial is approached.
The deciduous trees give way to the ever-
greens such as fir and pine with eventually
birch becoming abundant once more.

One interesting feature of this section, in
the early to mid Hoxnian is that there is a
marked period of deforestation with grasses
becoming suddenly more common. Artefacts of
man have been found within this zone but it
is not possible to say whether it was man
that caused the deforestation or if deforestation allowed the incoming of man.

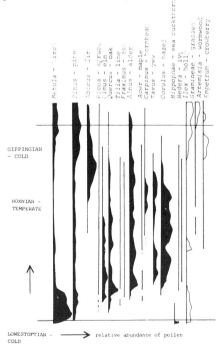

Fig. 7.4 Pollen diagram from the Hoxnian
interglacial lake deposit at Marks Tey, Essex.
Values expressed as percentages of total land
pollen.

Spores

Fossil spores are usually divided into two
major groups on a size basis. Megaspores are
those of over 200 microns in diameter and
miospores of less than 200 microns in
diameter.

Miospores are rather difficult to extract and
study but it is possible to break down a
sample of coal relatively easily using a
powerful oxidising agent to leave a residue
containing megaspores. These can be studied

quite easily with equipment available in schools. The detailed method for extraction of these megaspores is given in the appendix.

A number of detailed analyses of the spore and pollen content of coal have been carried out in order to try to establish the conditions and nature of vegetation associated with the coal deposits of the Upper Carboniferous. Interpretation is based upon relating the different types of spores and pollen to their parent vegetation.

One study carried out by A.H.V. Smith (1961) on some Westphalian coal seams of Yorkshire recognised a number of distinct miospore assemblages. Each of these spore assemblages reflects the character of the vegetation growing on or near the coal swamps at the time, and each assemblage was linked with a particular type of coal.

The assemblages were grouped into four categories on the basis of the dominant species. These were the Lycospore phase, Transition phase, Densospore phase and Incursion phase.

These phases often occur in a simple sequence with the the coal seams i.e. Lycospore-Transition-Densospore-Transition-Lycospore. However, changes do occur with parts of the cycle being repeated or even missing in different seams. Differences may even arise within the same seam at different localities which is a reflection of the variation that exists in the environmental conditions. The Incursion phase varies in position in the sequence coming either between the Lycospore and Transition phase or between the Densospore and Transition phase. This phase represents a period when the coal swamps became flooded.

The thickness of the coal associated with each phase may be only 2-3cm or it may be over 1m thick. With the Lycospore phase the coal is composed of vitrite and clarite which give it a bright appearance. In contrast the Densospore phase is mainly composed of durite which gives it a rather dull look. As one might expect, the composition of the Transition phase varies depending upon which other phase is being approached. The Incursion

phase contains fusinite which has a charcoal-like appearance.

The nature of the coal is controlled by the amount of aerobic decomposition to which the peats from which they were formed were subjected. Clarite and vitrite are thought to have been produced under anaerobic conditions whilst durite, which is made up of the most resistant parts of the plants is thought to be the result of more aerobic decomposition. Some people believe that the charcoal-like nature of the fusinite may be attributed to forest fires although it could also be the result of other processes.

Smith was able to use the information from the spore content and type of coal to build up a picture of the changing environments of the Coal Measure swamps.

1. Lycospore phase. This represents the initial stage of peat formation under a shallow covering of more or less stagnant waters giving anaerobic conditions. These conditions may have persisted for a long time if the balance between peat formation and subsidence was maintained. The vegetation at this stage was forest, *Lycospora* (Fig. 7.5) being a dominant spore. This comes from arborescent lycopods, such as *Lepidodendron* which grew to a height of 30 m (Fig. 7.6).

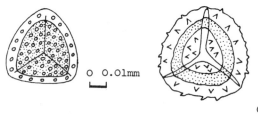

Lycospora   Densosporites

Fig. 7.5 Lycospora *and* Densosporites.

2. Transition phase. The decomposition in this phase became more aerobic as there was a drop in the level of the groundwater. This was either related to a fall in sea level or because the level of the bog rose and more peat accumulated in the humid climate. These changes, together with a possible change in

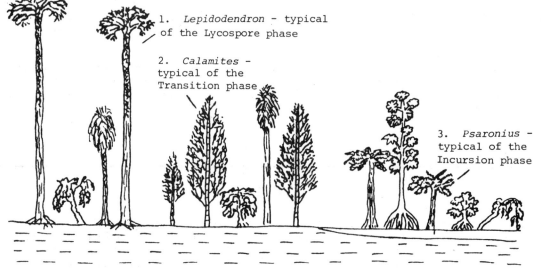

1. *Lepidodendron* - typical of the Lycospore phase

2. *Calamites* - typical of the Transition phase

3. *Psaronius* - typical of the Incursion phase

Fig. 7.6 Reconstruction of some of the trees of a Coal Measures swamp.

climate, resulted in the gradual replacement of the forest by more open vegetation. Spores of herbaceous lycopods, horsetails (e.g. *Calamites*) and ferns are found in this phase (Fig. 7.6).

3. Densospore phase. The interpretation of this is not quite as well understood as the other phases. *Densosporites* (7.5) is the most common spore in this phase and the content of the coal here shows marked decay, which is probably a result of aerobic decomposition. This could be linked to a lack of surface water which may possibly be caused by climate changes. This phase will be followed by a return to the transition phase.

4. Incursion phase. This is caused when the normal events are interrupted by flood water which has a catastrophic effect on the existing vegetation. The spore assemblage is rich in species with herbaceous plants being dominant. Following this flooding the pattern becomes re-established.

It is not always possible to be totally accurate when using these spores to reconstruct the Carboniferous vegetation as it is no easy matter to link a particular spore type to the rest of the plant. Nevertheless, much valuable information can be gained about the changing nature of the coal measure swamps using this technique.

Conclusion

We have seen in looking at these three groups of microfossils - foraminifera, spores and pollen - how useful they can be in palaeo-ecological reconstructions. But it is their size and consequent ease of recovery from cores and chippings from boreholes which make them so invaluable in these situations where evidence from macrofossils is exceedingly unlikely. Indeed, over recent years a great deal of research has gone into both the palaeoecological and the stratigraphic applications of microfossils.

Appendix

Extraction of spores from coal

With very simple equipment, miospores are difficult to extract and study easily but it is relatively easy to obtain your own sample of megaspores from coal.

About 10 grams of coal are required and this is broken down into small fragments and placed in a clean glass flask. An oxidising agent then has to be added to cover the sample. This is Schulze's solution (cold concentrated nitric acid: saturated potassium chlorate (3:1). Great care must be taken over this part of the process. Protective clothing and goggles need to be worn and it must be carried out in a fume cupboard as the reaction may be violent. The flask can be placed in a large beaker of cold water. It will be sufficient to leave most samples overnight but some fresh samples, such as those from underground workings may need as long as 72 hours. However, with a severely weathered sample this stage may even be missed out.

Pour off the solution taking care not to lose any of the sample. Add distilled water and keep repeating the dilution process until as much of the solution as possible has been removed. Next add 250 ml of potassium hydroxide solution (5-10%) and stir gently with a glass rod. The mixture will now have turned dark brown. Leave the sample for about 4 hours, which will give enough time for the humic substances in the coal to be dissolved.

Now wash the sample through a fine sieve and transfer to a small glass beaker with a fine jet of water. This residue should contain megaspores along with some other plant debris. Take a small amount of the sample and place it in a flat bottomed dish under a microscope using reflected light. A magnification of about X50 is required. The megaspores should be yellow brown. If they are brown or black further oxidation is required, or if the water turns brown during examination of the specimen some more alkali (KOH) should be added followed by further washing with distilled water.

The technique of separating the spores from the debris requires patience and a steady hand using a fine brush and a needle. The spores can be placed on a glass slide for examination in either transmitted or reflected light depending upon the type of microscope available. The smaller thin walled spores are better studied in transmitted light, but for the large thick walled types reflected light is better.

They can be stored in a glass tube containing distilled water plus a drop of hydrochloric acid or a more permanent mount can be made on a glass slide. A water mounting medium or glycerine jelly may be used and the cover slip placed over the spores. These types of mounts are not permanent, as with time they will dry out. Canada Balsam must be used for a permanent mount but the spores must be dehydrated in alcohol first.

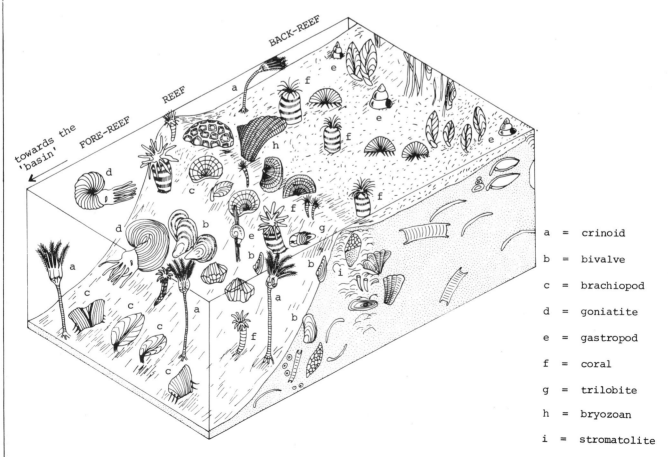

a = crinoid

b = bivalve

c = brachiopod

d = goniatite

e = gastropod

f = coral

g = trilobite

h = bryozoan

i = stromatolite

Fig. A.1 Artist's reconstruction of a Lower Carboniferous reef complex.
(Not to scale)   The reef probably had a relief of 100 m or so above the adjoining 'basin'

Further Reading

Black, R.M.
The Elements of Palaeontology. Cambridge
University Press. 1970. Standard introduction
to the morphology and geological occurrence
of the main groups of organisms.

Clarkson, E.N.K.
Invertebrate Palaeontology and Evolution.
Allen & Unwin. 1979. An advanced but
comprehensive and up to date reference book.

Kirkaldy, J.F.
Fossils in Colour. Blandford Press. 4th ed.
1975. General introductory text on fossils -
their classification, occurrence and uses.

Laporte, L.F.
Ancient Environments. Prentice-Hall, Inc.
1968. Deals with the main sedimentary
environments, past and present.

McKerrow, W.S.
The Ecology of Fossils. Duckworth. 1978.
Contains reconstructions in block diagram and
map form of most of the major ecosystems of
the British statigraphic record.

Raup, D.M. & Stanley, S.M.
Principles of Palaeontology. Freeman. 2nd ed.
1978. An advanced but valuable reference
book giving the main principles of the subject.

Bibliographical References

Ager D.V.
'The teaching of palaeontology', *Geology
Teaching*, vol 1 no 3 1976.

Ericson, Ewing and Wollin
'The Pleistocene-Pliocene boundary in deep
sea cores', *Science*, vol 139 no 3556, pp
727 to 737, 1963.

Hallam A.
'Observations on the palaeoecology and ammonite sequence of the Frodingham Ironstone
(Lower Jurassic).*Palaeontology*, vol 6 pp
554-574,1963.

Higgins A.C. and Spinner E.G.
'Techniques for the extraction of selected
microfossils', *Geology - Journal of the
Association of Teachers of Geology*, vol 1,
pp 12-28, 1969.

Ireland R.J., Pollard J.E., Steel R.J. and
Thompson D.B.
'Intertidal sediments and trace fossils from
the Waterstones at Daresbury, Cheshire',
*Proceedings of the Yorkshire Geological
Society*, vol 41 pp 399-436, 1978.

Smith A.H.V.
'The palaeoecology of Carboniferous peats
based on the miospores and petrology of
bituminous coals. *Proceedings of the Yorkshire
Geological Society*, vol 33, pp 423-474. 1962.

Fossils & Time

Contents

Acknowledgements

The authors wish to acknowledge the valuable assistance of the following in the preparation of this Unit: Dr Richard Porter for the photographs in Section 5 and for his comments on that section; Dr Frank Spode for his critical reading of the manuscript and for help with sources of information; Dr Hugh Torrens for permission to reproduce part of a coal company prospectus; Mrs J. Kay and Mrs S.E. Ross for translating unreadable scrawl into neat typescript.

The authors alone are responsible for any inadvertent errors.

We are grateful to the Institute of Geological Sciences for permission to reproduce Fig. 4.6.
The following figures have been based on illustrations from the sources indicated:
Fig. 1.4 Eicher, D.L. *Geological Time*, Prentice Hall International; Fig. 1.5 Raup, D.M. and Stanley, S.M. *Principles of Palaeontology*, Freeman and Co; Fig. 2.2 Donovan, D.T. *Stratigraphy: an Introduction to Principles*, Wiley; Fig. 2.3 Clarkson, E.N.K. *Invertebrate Palaeontology and Evolution*, Allen and Unwin; Fig. 2.7 cover of Institute of Geological Sciences Map, *Llandrindod Wells Ordovician Inlier*; Fig. 2.10 ed. Basset, D. *The Ordovician System Symposium*, University of Wales Press; Fig. 2.11 Moseley, F. *The Geology of the Lake District*, Yorkshire Geological Society; Fig. 3.3 Kellaway, G.A. and Welsh, F.B.A. *British Regional Geology, Bristol and Gloucester District*, Institute of Geological Sciences; Fig. 3.8 Ramsbottom, W.H.C. *Transgressions and Regressions in the Dinantian*, Yorkshire Geological Society; Fig. 3.9 as Fig. 3.8; Fig. 4.3 Anderton, Bridges, Leader, Sellwood, *Dynamic Stratigraphy*, Allen and Unwin; Fig. 4.5 Calver, M.A. 'Westphalian of Britain'. *Compte Rendu, 6th International Congress on the Stratigraphy and Geology of the Carboniferous, Vol 1*, Sheffield.

Foreword

The use of fossils in stratigraphy is an old established technique, yet in spite of the advent of radioisotopic dating, it remains of immense value, both to the economic geologist and to the academic. With a little experience, fossils can be used by relative beginners in geology to assign rocks found during their fieldwork to the appropriate part of the geological column. Fossils and Time includes a number of case studies to illustrate the different principles which are involved. Other methods of establishing geological dates are covered in different Units. It is assumed throughout that the reader has a working knowledge of the nature and morphology of the common invertebrate groups of fossils and a little familiarity with fossils from the Plant Kingdom.

The case studies you will read of have been chosen to illustrate some of the variety of situations where fossils can be of great value in stratigraphy. Some of the studies include exercises and the reader will gain most from the Unit if these are attempted as they arise, rather than glossing over them, or cheating by looking up our answers in the Appendix! It is worth stressing that the authors' answers are not the only ones possible; they are included merely as a guide to the type of interpretations which may be reached.

1

Introduction

Historical Development of the Subject

For hundreds of years fossils have been collected from the rocks. At first they were merely objects of interest and curiosity, as it was not fully appreciated what they were. Even the word 'fossil', derived from a Latin word meaning 'dug up', was originally applied to a whole range of items, some organic in origin, but many of an inorganic nature. Later, the word became restricted to the remains or traces of once living organisms, which were regarded as having been 'trapped' in the rocks. Just how and why they got there was still something of a mystery and how they fitted in with prevailing religious ideas on the Creation was a matter of contentious debate. It was not until the nineteenth century, as a result of work by people such as Darwin, that their significance as evidence of stages in the evolution of life was at least tentatively understood.

However, even before these ideas of evolution had been voiced to a sceptical and hostile world of the mid nineteenth century, fossils had proved useful to geologists. Extensive collecting, description and classifying of fossils had gone on for a long time, during which, incidentally, many clergymen had played a key role, so the various groups of fossils and the diversity amongst these groups was quite well known.

In a number of places, at about the same time, people started to notice that there was a pattern associated with the occurrence of these fossils. One of these workers was William Smith who has been called the 'father of British stratigraphy'. William Smith was not a geologist by training, but a civil engineer whose work building canals and bridges took him to a number of different parts of the country. In his work, directing the construction of canals, for example, many of the channels for which he was responsible were cut through several layers of sedimentary rocks. He soon noticed that the fossils contained within these layers of rocks were distinctive for each level. In various parts of the country he was able to recognise rocks of a similar age because of the same assemblage of fossils that they contained. He was even able to tell apart rocks of different ages, but of similar lithologies, by using their different fossil content. He then used these principles to produce, in 1815, the first geological map of England, Wales and parts of Scotland. These early maps by Smith coincide remarkably closely with some of the boundaries mapped today. He was working well before the stratigraphic column was divided up into the now familiar systems (Cambrian, Ordovician,

etc.), but some of his divisions such as the 'Mountain Limestone' compare quite well with the Carboniferous Limestone we talk of today.

A little later, in France, Georges Cuvier also used fossils to divide up the strata in the Paris Basin. Unlike previous studies, which used mainly marine invertebrates, Cuvier concentrated on using terrestrial vertebrates in the Tertiary parts of the succession. Many fossils of marine invertebrates that had no known living counterparts had previously been discovered in rocks of all ages. The explanation given for this was that the living counterparts had not been discovered as yet, since marine biology was then in its infancy. With the terrestrial vertebrates that Cuvier discovered there were also cases where there were no modern equivalents. He came to the conclusion, as it was highly unlikely that quite large land animals could remain undiscovered, that there must have been 'extinction' of animals.

He also noticed that the younger rocks contained fossils of creatures which were more like those of the present day. Cuvier found many abrupt faunal breaks in the succession in the Paris Basin, some of which coincided with lithological changes such as a conglomerate bed. He suggested, therefore, that these breaks were the result of a series of catastrophes that had destroyed the fauna which then became extinct. After each break new fauna were created, each time becoming more complex and more like modern life forms.

Even though Cuvier's ideas of a series of catastrophes, culminating with Noah's Flood, were eventually discounted, many geologists in different parts of the world were able to use the ideas of both Smith and Cuvier until, gradually, a series of geological systems became recognised. These systems 'evolved' over a number of years (1760-1891), often after much debate, which in some cases is still going on, about where the boundaries between the systems should be drawn.

Most of the geological systems, which have now become established in the geological record, were first defined in Europe. Each system has some rock outcrops known as the type section (or stratotype) on which it is defined and other rocks of equivalent ages are correlated with the type section. The area in which the type section or sections have become defined is known as the type area. In many cases the original type section has proved unsuccessful for correlation purposes outside the area, because it has been unfossiliferous, or because parts of the sequence are missing, or because it was defined on a facies that was not particularly widespread. So the type sections sometimes have to be redefined from a more suitable locality. Table 1.1 shows how the names of the systems originated. The equivalent time interval is a period.

TABLE 1.1 THE STRATIGRAPHIC COLUMN SHOWING THE ORIGIN OF THE NAMES.

ERA	PERIOD	EPOCH		
CAINOZOIC (new life)	QUATERNARY	Third and Fourth, following on from the Palaeozoic (Primary) and Mesozoic (Secondary) Eras	Holocene	–	wholly recent
			Pleistocene	–	almost recent
	TERTIARY		Pliocene	–	more recent
			Miocene	–	less recent
			Oligocene	–	little recent
			Eocene	–	dawn of recent
			Palaeocene	–	ancient recent
MESOZOIC (middle life)	CRETACEOUS	–	Derived from 'creta' the latin word for chalk as this is the distinctive rock in the upper part of the system in the Paris Basin and England.		
	JURASSIC	–	Named from the Jura Mountains in France which contain strata of this age.		
	TRIASSIC	–	In Germany, where this period was first named in 1834, there is a clear three-fold division.		
PALAEOZOIC (old life)	PERMIAN	–	In 1841 a British geologist called Murchison named this after Perm in Russia where there was a thick sequence of limestones containing distinctive fossils.		
	CARBONIFEROUS	–	There is a widespread occurence of carbon-bearing rocks (coal) in this system. Even in 1822 when it was first named by Conybeare and Phillips, they realised that it was the distinctive fossils rather than the lithology which were important. In North America the terms Mississippian and Pennsylvanian (after the two regions) are used for the Lower and Upper Carboniferous respectively.		
	DEVONIAN	–	Named after Devon, which was the original type area, in 1840 by Murchison and another British geologist Sedgwick. They were later to have a dispute over the naming of periods in Wales.		
	SILURIAN	–	All these periods were named from Wales. Cambria was the Roman name for Wales and the Ordovices and Silures were ancient Celtic tribes of Wales. Sedgwick and Murchison worked on the strata in Wales at the same time, Sedgwick starting at the bottom of the section and working up, and Murchison working down from the top of the sequence. They called their systems Cambrian and Silurian respectively. An argument ensued when it was realised that their systems overlapped and this was not resolved until 1879 when Lapworth proposed the name Ordovician to cover the disputed interval.		
	ORDOVICIAN	–			
	CAMBRIAN	–			
	PRECAMBRIAN				

Principles of Zoning

As the various systems were being defined using fossils, it soon became clear that certain fossils were more useful than others for indicating a relative age for the rocks. Obviously, if a species exists for a long period of time without change, or with only minor changes, it will not be as useful for assigning an age to the rocks as a species that has only a short time range. Figure 1.1 illustrates at what point in time members of each of the major fossil groups have proved useful for zoning.

Fossils which are of value in dating the rocks are known as stratigraphic or zone fossils. To be a successful stratigraphic fossil, it will need to fulfil all or most of the following criteria. The species needs:

(i) To be easily recognisable and distinguishable from its immediate ancestors and descendants.
(ii) To have existed for as short a period of time as possible.
(iii) To be preserved easily and for the distinctive characteristics of the species to be recognised after fossilization.
(iv) To be geographically widespread so that it can be used in a number of different areas.
(v) To be abundant so as to increase the chances of being found.

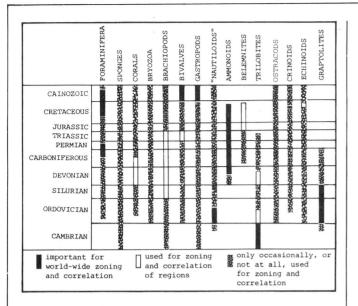

Fig. 1.1 Relative value of major groups of marine invertebrates during the Phanerozoic.

Legend for Fig 1.1:
- ■ important for world-wide zoning and correlation
- □ used for zoning and correlation of regions
- ▨ only occasionally, or not at all, used for zoning and correlation

(vi) To be present in a number of different facies.

It is rare to have a fossil that meets all the above requirements, but free swimming or floating forms such as ammonites and graptolites have proved to be very successful because they show the regular changes needed and their life style led to a wide distribution of individuals.

Species can suddenly appear in a sequence of rocks for a number of different reasons. Changing environments can lead to a species or a number of species new to an area being found, i.e. the occurrence is related to the new facies. In different areas these species may come in at different times, because conditions change at different rates, that is they are diachronous. These assemblage zones, as they are known, are not too useful to stratigraphers as they are not totally dependent on time, but they may be more use in palaeoecological studies. (See the Unit Palaeoecology)

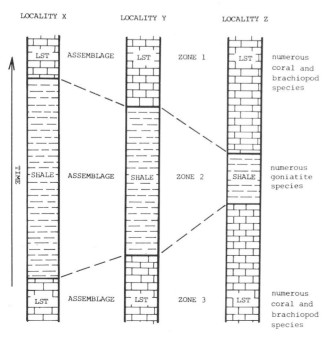

Fig. 1.2 Assemblage zones.

Figure 1.2 illustrates 3 sections within an area. Each shows the same sequence of rocks containing the same assemblage zones but the changes from one zone to another did not happen at the same time in each of the localities. At locality X the conditions suitable for the goniatites lasted for considerably longer than at locality Z. Therefore, these assemblage zones cannot be used for accurate time correlation.

It is more useful to the stratigrapher when a sequence of species can be seen. In the most straightforward case, one species may evolve into the next species which in turn gives way to a third species. This happens when the species are part of a phyletic lineage and will give rise to range zones (Fig. 1.3). Not all range zones are as straightforward as the case illustrated. Apart from giving way to a descendant there are many other reasons why a species may appear or disappear in a particular area. Geographical barriers may break down, allowing migration into an area, or there may be subtle changes in the environment allowing in new species.

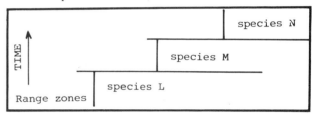

Fig. 1.3 Range zones.

The sudden disappearance of a species which does not leave a direct descendant may also be due to a number of causes. Environmental changes could again have been responsible. This could lead to problems of correlation over wide areas, as species could well be exterminated in different places at different times, and so it does not give a synchronous event.

The most accurate drawing of zone boundaries will be achieved by using the overlapping ranges of a number of species, as this reduces the possibility of a range of a particular species being abnormally extended or reduced at a particular locality. These overlapping ranges are known as concurrent range zones (Fig. 1.4).

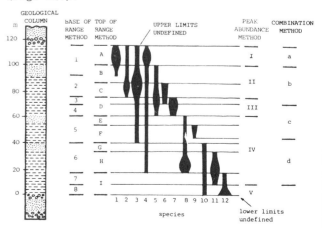

Fig. 1.4 Four methods of recognising concurrent range zones in a single section. Width of lines indicates the relative abundance of specimens.

By using concurrent range zones, the problem of species being closely related to facies (as in assemblage zones) is lessened as species can be selected that occur in different facies. In fact, some of the species commonly found in a particular zone may never be found in the same beds.

The concurrent range zone will be named after one of the species within it, usually after the most characteristic or abundant species, but it need not necessarily be any more significant than any of the other species within the zone. Indeed, the named species need not even be confined to the zone.

Even by using concurrent range zones involving a number of species there may still be problems about transferring the scheme from one province to another, because extinctions occur in different environments at different times. Some of these problems are illustrated in a later section of this Unit.

With overlapping time ranges, the problem is whether to draw the boundary between zones on the incoming of a new species or the extinction of a declining one. Traditionally, many zones have been drawn on the incoming of a new species although there may be merits in using extinction or a combination of both. For instance, when details of the subsurface geology are revealed by drilling, it is convenient to use the upper limits of a species.

This is partly as a result of approaching the section in reverse order, but also to minimise the chances of error caused by specimens falling down from the sides of the well and being brought up from a lower level than the one in which they occur.

Some attempts have been made to use the relative abundance of certain species rather than the end points of their range alone. The zone boundaries have not been delimited on the arrival or disappearance of the species but by its time of most abundance. Such peak abundance zones do work in some localities and an example of where they have been used is given in Section 3 of this Unit. Their use is obviously full of difficulties as the time of peak abundance may well be an environmentally controlled factor rather than a time-significant factor. Nevertheless, they may be used successfully, especially over small, restricted areas.

Figure 1.4 shows the various ways concurrent range zones may be used for a single section. Depending upon which method is selected, the number of zones varies from 4-9. Other combinations of the various methods will produce further ideas for dividing up the sequence. Whichever method is chosen largely depends upon the particular area of study. The later sections of this Unit illustrate a wide variety of techniques used in different circumstances.

Problems of Defining Species

Whichever method of zoning is chosen, it is crucial to be able to distinguish between species. The biological definition of a species uses the fact that members of the same species can interbreed, a method which is not available to palaeontologists! Instead of dealing with biospecies, palaeontologists usually have to rely on morphospecies or 'form species'. That is, specimens are put in the same species if they look alike. Often in a particular sequence there is no problem in deciding whether or not there are two species when the species are sufficiently distinctive, but where a continuous phyletic lineage is used, even though the top and bottom members of the lineage are clearly in different species, it is often very difficult indeed to draw the dividing line when there has been gradual change.

Two contrasting schools of thought exist: those who cannot tolerate much change in the morphology and tend to 'split' the phyletic lineage into many different species and those who tend to allow quite a lot of variation within the species and 'lump' much of the phyletic lineage into a few species.

There are fewer problems where species are not part of the phyletic lineage as there may be marked differences between the incoming and existing species; indeed they need not necessarily be related at all.

Micraster, an Upper Cretaceous, irregular echinoid, provides an example of where a phyletic lineage has been divided into a number of species and it illustrates some of the problems that arise when making these divisions.

Micraster is found in the Middle and Upper Chalk of the South of England. The character of the chalk does not vary much throughout its thickness and this probably indicates that the environment was uniform. The well documented changes which Micraster shows have been interpreted as a response to a changing mode of life, rather than environments, and the nature of these changes suggests that Micraster increased its depth of burrowing.

Figure 1.5 illustrates the major differences between Micraster corbovis and Micraster coranguinum which represent two extremes in the sequence. These changes are:

(a) Increased ornamentation around the respiratory tube feet indicating that there were more cilia to improve the flow of water throughout the burrow.
(b) The anterior groove leading to the mouth is deepened and a lip develops below the mouth which also moves forward. This improved food gathering.
(c) A broader fasciole (area of ciliated spines) develops beneath the anus which improved sanitation.
(d) There were more and stronger spines on

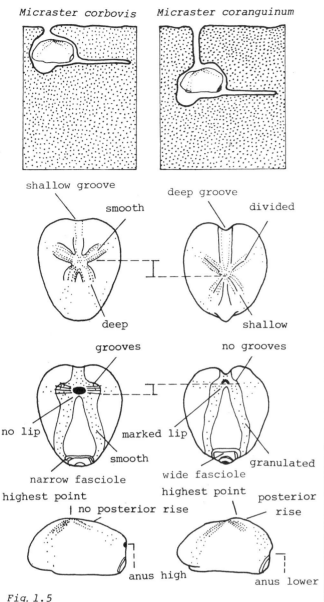

Micraster corbovis   *Micraster coranguinum*

shallow groove

deep groove

smooth   divided

deep   shallow

grooves   no grooves

no lip   marked lip

smooth   granulated

narrow fasciole   wide fasciole

highest point   highest point

| no posterior rise   posterior rise

anus high   anus lower

Fig. 1.5

the oral surface indicated by the tubercles present, which would have been used in digging.

(e) The test became broader and its highest point shifted towards the posterior as it became more streamlined.

All these developments would have aided *Micraster* to burrow deeper.

The original *Micraster* species was *M. leskei*, which was a shallower burrower, and this gave rise to another shallow burrower *M. corbovis* and one which started to burrow deeper in the sediment of the sea bed, *M. cortestudinarium*. There are no clear-cut differences between these two species as the changing life style developed gradually over time and many specimens recovered from the chalk of the south of England fall between the two types. The situation is further complicated in the Senonian stage where *M. senonensis*, another shallow burrower, enters the region after it had probably been geographically isolated. *M. coranguinum* had by this time developed from *M. cortestudinarium*, although the division between the two species is completely arbitrary. However, intermediate forms between *M. senonensis* and *M. coranguinum* suggest that they were not sufficiently genetically different to prevent interbreeding.

Figure 1.6 summarises the relationships between the 5 different species. There are obvious differences between say, *M. corbovis* and *M. coranguinum* and between *M. coranguinum* and *M. senonensis*, but between the extremes a series of gradual changes with time is represented. Just where to draw the line between the various species is sometimes a matter of personal opinion and in any case these divisions are artificial, as the changes between the species were a continuous process.

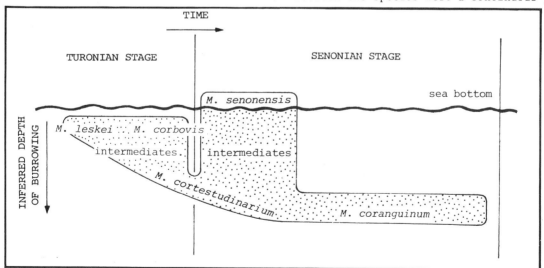

Fig. 1.6 Life habit changes in the evolution of Micraster in the Cretaceous of the south of England.

2

Graptolites and Trilobites in the Ordovician

Introduction

Much of northern and western Britain is composed of rocks of Ordovician age. Considerable diversity in lithology is exhibited from region to region, not only in the sedimentary rocks but also in associated volcanic sequences and igneous intrusions. Correlation between these Ordovician rocks is complicated by several factors, notably the destruction, by metamorphism, of the fossils which once existed in some of the sediments and the absence of fossils in the first place from volcanic rocks. Even where sedimentary rocks have remained relatively undeformed, there must have been considerable differences in their depositional environments from place to place. This is reflected in their fossil content. Thus, sandstones, limestones and the lighter coloured shales may well contain a diverse fauna representative of benthonic organisms which needed well-oxygenated and sunlit waters in which to thrive. Such faunas include trilobites, brachiopods, bivalves and crinoids and are usually indicative of the shelly facies or shelf facies of the Ordovician.

By contrast, many sequences of the same age consist of dark shales, some of them with inclusions of pyrite, which were produced from bacterial concentration of sulphides under anaerobic conditions on the sea bed. Such an environment can hardly have supported benthonic life and the main fossils to be found in the dark shale sequences are graptolites, which must have lived in the surface layers of the sea and only drifted down to the sea bed after their death. In some localities, graptolite fossils smother the lamination planes of the shales and it is clear that there could not have been any predators on the contemporary sea bed, or the graptolites would not have been preserved intact. Such shales constitute part of the graptolitic facies or basin facies. Other basin facies rocks consist of great thicknesses of greywackes, many of which are poorly fossiliferous.

This diversity in the Ordovician System poses obvious problems for time-correlation between the basin and shelf facies, but fortunately there are some areas where there is an overlap between them. Some of the shelf organisms were washed onto the stagnant basin floors and some stray graptolites survived the higher energy levels of the shelf.

Figure 2.1 summarises the facies and faunal contrasts in Wales and the Welsh Borderlands. A similar situation existed in the Southern Uplands. It should be noted that the 'basin' areas were not necessarily very deep water, as implied in the 'classic' diagram below. This section will examine the relative value of graptolites and trilobites as zonal fossils and demonstrate how a stratigraphic sequence may be established.

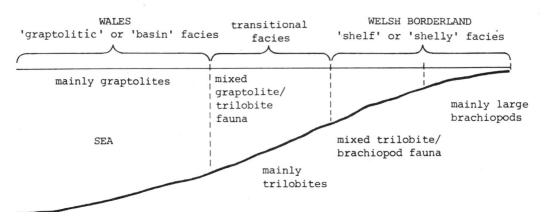

Fig. 2.1 Diagrammatic section of the Welsh area during the deposition of the rocks of the Ordovician.

Graptolite Zones

Suitability as zonal indices

The basic and often quoted requirements for a good zone fossil have been explained in Section 1. In many respects, the graptoloid graptolites fulfil these very well, although some cautionary words are necessary:

(a) They were mostly planktonic, probably floating at the mercy of ocean currents. Figure 2.2 shows how widespread are the local-

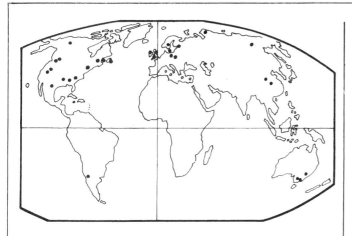

Fig. 2.2 *Distribution of the Ordovician graptolite* Nemagraptus gracilis.

ities where one particular species, *Nemagraptus gracilis* is found. However, it must not be forgotten that the continents have moved since Ordovician times. According to at least one reconstruction of the continents in the Ordovician, all the *Nemagraptus gracilis* localities would have then been in tropical or sub-tropical waters. Nonetheless the wide distribution is still quite impressive.
(b) Numerous specimens are available. This is true of certain well-known localities where lamination planes are thickly plastered with graptolites, but in other areas they seem to be few and far between. Perhaps some of the structural problems posed by the Skiddaw Group in the Lake District would be more readily solved if only a few more graptolites were present!

(c) The preservation of graptolite specimens is normally sufficiently good for them to be readily identified. This is often true but most specimens are the flattened remains of once three-dimensional objects and some distortion may have occurred.
(d) They evolved rapidly along distinct lineages. Figure 2.3 shows some of the main sequences of graptolites and demonstrates the marked distinctions between them. In general terms, the older forms have more stipes (branches) than younger ones. There is a tendency for the stipes to be pendant (to hang down) in the earlier forms and to become scandent (bent back) with the passage of time. The single stiped Monograptids developed early in the Silurian, although this did not coincide with the extinction of all the Ordovician forms. By the standards of the lower Palaeozoic, the refinement in dating which can be achieved using graptolites is very good, each major zone representing an average of four million years. However this is not as precise as the average 900,000 years per zone obtainable with ammonites for the Jurassic. Developments in micropalaeontology, too, have resulted in even more accurate zoning in some parts of the geological column.
(e) Graptolites should, in theory, have been free of facies restrictions by virtue of their pelagic mode of life. However, because of their frailty, they are rarely found in rocks of higher energy origin than siltstone. Even where low-energy environments did occur in shelf seas, the rapid oxidation of their protein-based skeleton and the activity of predators would have ensured the paucity of

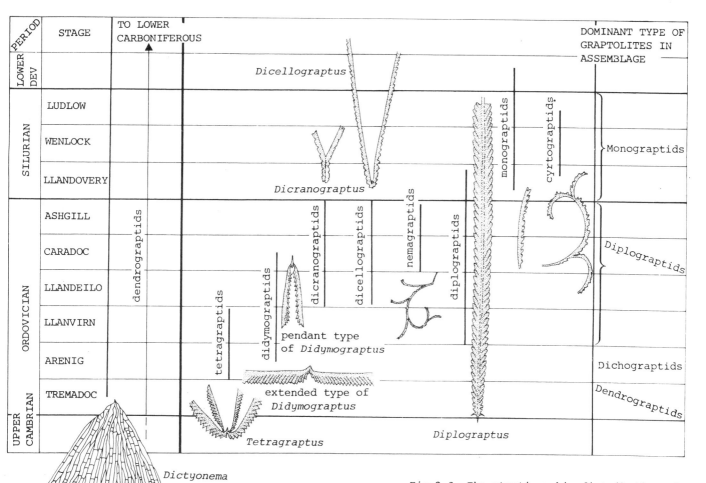

Fig. 2.3 *The stratigraphic distribution of some of the main groups of graptolites.*

graptolite remains in the fossil record. These factors explain why they are associated with the reducing conditions of the dark shale facies rather than any other environments.

An exercise in zoning part of the Ordovician

Graptolites were first effectively used in zoning by Charles Lapworth. In the late nineteenth century he established a series of graptolite zones in the Ordovician and employed them to unravel some of the complexities of the geological structure of the Southern Uplands of Scotland. Thirty years or so later G.L. Elles and E.M.R. Wood extended the zonal system, mainly from their work in Wales. Elles' and Wood's zones are little changed today.

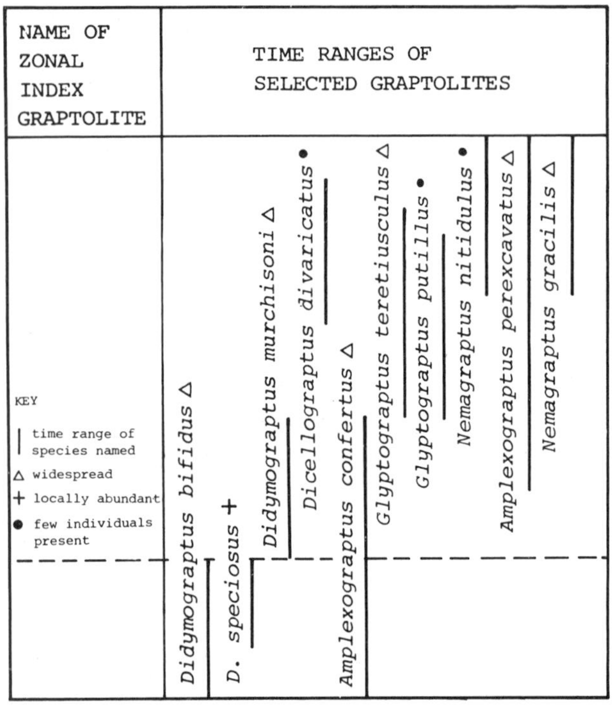

NAME OF ZONAL INDEX GRAPTOLITE	TIME RANGES OF SELECTED GRAPTOLITES

KEY

| time range of species named
△ widespread
+ locally abundant
● few individuals present

Didymograptus bifidus △
D. speciosus +
Didymograptus murchisoni △
Dicellograptus divaricatus ●
Amplexograptus confertus △
Glyptograptus teretiusculus △
Glyptograptus putillus |
Nemagraptus nitidulus ●
Amplexograptus perexcavatus △
Nemagraptus gracilis △

Fig. 2.4 Time ranges and relative abundance of selected graptolite species in part of the Ordovician.

In this exercise, try to imagine that you are in the position of these pioneer workers, having recorded a large number of species of graptolites from successively higher strata in a variety of localities. Study Fig. 2.4 which shows the time ranges of some of the graptolite species from the mid Ordovician, with particular reference to Wales. Try to choose the best graptolites to set up zones which are likely to be of widespread validity. The lowest zone has been done for you, although you should choose the best name for the zone from the fossils available. Remember the principles about the use of graptolites outlined earlier, and the following:

(a) Zones are usually of roughly even time span.
(b) In cases of overlap, take the incoming of a new species as the top of the lower zone, rather than the extinction of an existing species. (Fig. 2.5).

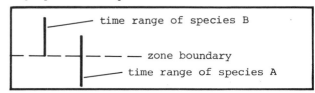

time range of species B
zone boundary
time range of species A

Fig. 2.5 Choice of zonal boundary when time ranges of species overlap.

(c) Although you may have named a zone after one fossil, you should always aim to use the concurrent ranges of several fossils to do the zoning in practice.

When you have made your choice, draw your zonal boundaries on a copy of the diagram, (Fig. 2.4) and choose a suitable name for each zone. In addition:

(a) Which species would be locally useful in providing 'sub-zone' information.
(b) Which zones are range zones and which are concurrent range zones, as defined in Section 1?

Try to resist the temptation to look in the Appendix, where the answers are given (Fig. A.1)!

Zones of the 'Shelly Facies'

The fossils of the shelly or shelf facies do not fulfil so many of the requirements for a zone fossil as those set out in the previous section. Most of the organisms were benthonic, such as trilobites, crinoids and brachiopods and cannot be expected to be as widespread as the graptolites. They are therefore of value as local indices only. Evolutionary changes are less marked, so they do not allow the zonal precision possible with the rapidly changing graptolites. Pelagic creatures such as nautiloids are often present in the shelf facies but these exhibit even less evolutionary change and so are of little value in zoning.

Nevertheless, shelly facies fossils may have to be used if graptolites are not preserved, and provided that an abundant fauna is available with a good overlap between the

ranges, a reasonably accurate attempt at zoning may be made. The value of such zoning is enhanced if it can be related to the standard graptolite divisions, based on areas where the shelly and graptolitic facies overlap.

A correlation chart giving the time ranges of selected trilobites in comparison to the graptolite zones is shown in Fig. 2.6. As an illustration of the use of the method, you should try working through the following problem:

Some years ago, builders at Llandrindod Wells in Powys, mid-Wales, (Fig. 2.7) were reworking some old quarry waste to use as 'fill' in a house-building programme. Permission was obtained to collect fossil specimens, some of which are shown in the photographs (Fig. 2.8).

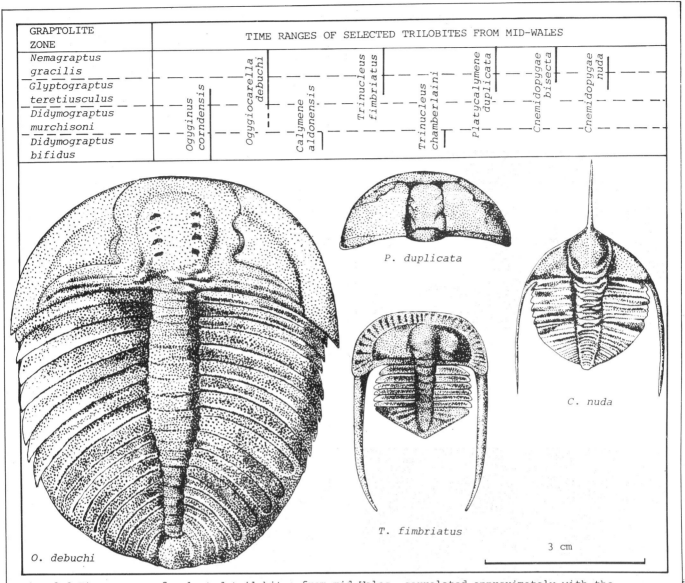

Fig. 2.6 Time ranges of selected trilobites from mid-Wales, correlated approximately with the standard graptolite zones.

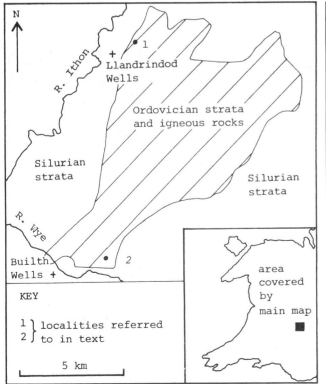

Fig. 2.7 Sketch map of the Ordovician inlier of the Builth Wells-Llandrindod Wells area, Powys, Wales

The rock in which they are preserved is a dark grey siltstone and, although the nearby quarry was originally worked for dolerite, there is no appreciable sign of thermal metamorphism in the waste siltstones. Abundant fossils were present, but in spite of an extensive search, no complete trilobites were discovered. The variety of species is indicated by the photographs of the recently collected specimens and in the faunal list on the next page, taken from the most complete description of the quarry, made in 1939.

Fig. 2.8 Fossils from Llandrindod Wells.

A. a thorax/pygidium, cephalon and part of the glabella of three different trilobites.

B. cephalon of *Platycalymene duplicata*

C. graptolites - long-ranging climacograptids

D. pygidium of *Trinucleus fimbriatus* and *Dicellograptus divaricatus*

E. cephalon of *Teleomarolithus intermedius*

F. part of a trinucleid cephalon on a fragment of a large asaphid trilobite

G. portions of *Ogygiocarella debuchi* (Scale bar is 5 cm total length)

In addition, Elles found several species of graptolites, but the recent work revealed only a number of climacograptids of indeterminate species and one specimen of *Dicellograptus divaricatus*.

On the basis of the above list and the photographs, try to answer the following:

(a) Use the ranges of the trilobites in the assemblage to define as closely as possible the zone from which the material came. Use the name of the appropriate graptolite index, even though it is not present in this collection.
(b) Do you consider that the waste material was produced by the one quarry, or was some of it brought from elsewhere?
(c) Comment on the state of preservation of the fossils.
(d) Using the answers to the above, the lithology of the host rock and your own knowledge of the usual habitats inferred for trilobites when they were alive, try to interpret the environment under which these siltstones were deposited in the Ordovician.

(e) The sketch in Fig. 2.9 shows some of the fauna found in mudstones near Builth Wells, about 11 km south of Llandrindod Wells. To what extent can you correlate the deposits of the two localities? Answers will be found in the Appendix.

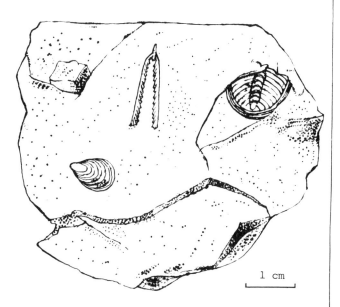

1 cm

Fig. 2.9 Ordovician mudstone containing *Didymograptus bifidus,* a trilobite pygidium (probably of *Ogyginus corndensis)* and an inarticulate brachiopod. A well-developed cube of pyrite is also present.

Faunal Provinces in the Ordovician

Although trilobites have been portrayed as the 'poor relation' among zonal indices for the Ordovician, there is a rather unexpected bonus to be derived from the very fact that they lived on the sea bed. When the benthonic 'shelly' faunas of Wales are compared with those of the same age and facies in Scotland, many marked differences in species are observed. Indeed, the Welsh faunas are more similar in their biological affinities to those of Europe, the southern part of Newfoundland and the east coast states of the U.S.A. than to the much nearer Scottish ones. Likewise, Scottish trilobite species are akin to those of Norway, Greenland, northern Newfoundland and most of the rest of the U.S.A., excluding the east coast states. Areas which have such close similarities in their faunas are regarded as constituting a faunal province. The map (Fig. 2.10) illustrates the British parts of the two distinct faunal provinces referred to and includes part of a third one, the Baltic province, which is identified by some geologists. Facies contrasts are also shown.

Careful tracing of evolutionary lineages of the 'shelly' fauna of each province, notably the trilobites, has demonstrated a common ancestry, probably in the Lower Cambrian, followed by gradual diversification into the separate faunas in the Ordovician. By Silurian times the distinction became blurred and separate faunal provinces are no longer identifiable.

The interpretation of these observations has depended to some extent on current fashions in geological thought. So long as the continents were regarded as having remained in fixed positions, various physical barriers between Wales and Scotland had to be invoked, across which the shelf-dwelling organisms could not migrate. Suggestions ranged from land barriers to deep-sea trenches.

Now that plate tectonic theory has become so well substantiated it seems natural to try to interpret the past on a plate tectonic framework. Further evidence has come from palaeomagnetic studies and from the identification of certain suites of igneous and metamorphic

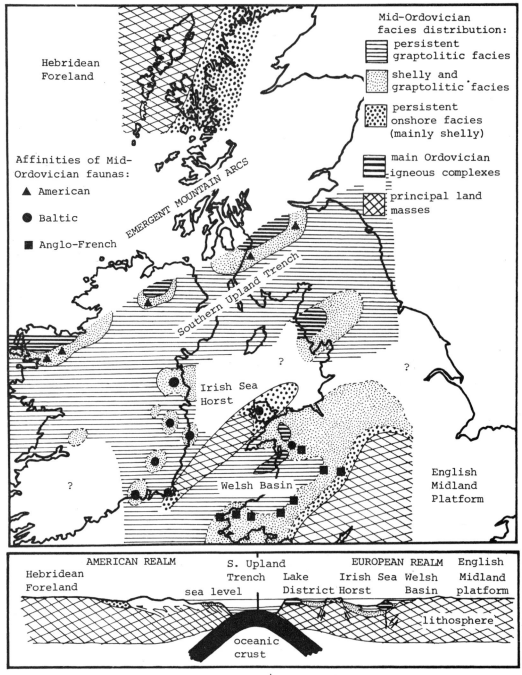

Mid-Ordovician
facies distribution:

persistent
graptolitic facies

shelly and
graptolitic facies

persistent
onshore facies
(mainly shelly)

main Ordovician
igneous complexes

principal land
masses

Affinities of Mid-
Ordovician faunas:

▲ American

● Baltic

■ Anglo-French

Hebridean
Foreland

EMERGENT MOUNTAIN ARCS

Southern Upland Trench

Irish Sea
Horst

Welsh Basin

English
Midland
Platform

?   ?   ?

AMERICAN REALM   EUROPEAN REALM   English
Hebridean   S. Upland   Midland
Foreland   Trench   Lake   Irish Sea Welsh   platform
          sea level   District Horst   Basin
                                                    lithosphere
          oceanic
          crust

Fig. 2.10 Mid Ordovician facies and faunal distribution within the Ordovician
successions of Britain and Ireland, with the section below showing the
inferred positions of converging plate margins.

rocks which, it is claimed, are characteristic of former destructive plate margins. It
is now suggested that there was once a whole
ocean (known to geologists as 'Iapetus')
separating England and Wales on the one side
from Scotland and most of North America on the
other. This began to open during Cambrian
times and was at its widest in the middle of
the Ordovician, when the separate provinces
of shelf faunas were most distinct (Fig. 2.11).
As well as the wide ocean itself, there were
probably two deep water trenches, making the
isolation of the two opposite shelves even
more complete.

Beneath Fig. 2.10 is a proposed plate tectonic framework. This interpretation is, of
course, subject to change as more information
becomes available, but at least it provides a
working hypothesis for the time being.

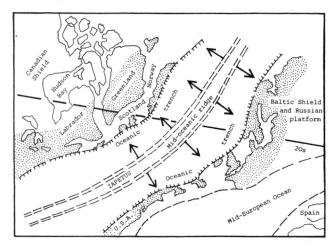

Fig. 2.11 'Iapetus Ocean' during the Lower
Ordovician.

3

Old and New Approaches to the Lower Carboniferous

Introduction

The preceding study has demonstrated the benefits of using pelagic organisms for zoning in preference to benthonic ones. Sometimes, however, fossils of usable pelagic creatures are completely absent from the geological record and benthonic animals have to be used, although they may be far from suitable. Because of the great economic importance of the rocks of the Carboniferous age, a disproportionate amount of effort has been devoted to establishing chronological sequences, to allow for correlation between the separate outcrops, mines and boreholes. A classic example was the work of A. Vaughan who set up a zonal system for the Lower Carboniferous of the Avon Gorge, using notoriously facies-controlled animals, namely corals and brachiopods.

Since the initial publication of Vaughan's paper, in 1905, his techniques have been widely employed by geologists, whilst at the same time they have been subjected to an extraordinary number of criticisms and modifications. Even Vaughan himself found it necessary to modify the system several times within the space of a few years. Although the system was in widespread use by the 1930s, the Geological Survey of that period temporarily abandoned it and reverted to lithological mapping alone, because it was felt that the corals and brachiopods were so facies-sensitive that misleading correlations would be made.

Nonetheless, Vaughan's method led to very important advances in Carboniferous stratigraphy and stimulated major attempts to correlate the Lower Carboniferous rocks of 'shelf-facies' throughout England and Wales. His technique also provides a useful example of principles which are not covered elsewhere in this Unit.

The Vaughan Zones

A summary of Vaughan's original zonal divisions with some of the fossils which he used is given in Fig. 3.1. The code letters for each zone are shown at the top of the columns. These represent the initial letters of the main zone fossils with alternatives where two fossils have the same letter or where the name has since been changed. Several generations of geologists have committed this sequence to memory through

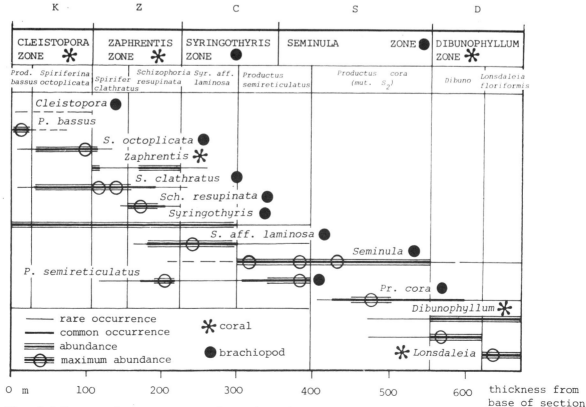

Fig. 3.1 Ranges of some genera and species of corals and brachiopods in the Carboniferous Limestone of the Avon Gorge, Bristol.

the use of the mnemonic, 'King Zog Cannot Sit Down!'

This zonal system was produced after much painstaking work by a conscientious and competent man, but it is all too easy to criticise. Thus:

(a) It did not prove possible to choose members of steadily evolving lineages as zone fossils; instead, the choice is more arbitrary and alternates between corals and brachiopods, depending upon which were commonly preserved.

(b) Some of the divisions were based upon a misunderstanding of the evolutionary history of the groups concerned.

(c) Although based on overlapping time ranges, like many other dating systems in the geological record, there is considerable variation in the abundance of specimens and hence in the likelihood of finding a particular species. For example - *Productus semireticulatus* has two periods of maximum abundance (acme) with a very lean time in between. One of these acmes is in the sub zone bearing its name, but the other is two sub zones earlier.

(d) Perhaps the biggest difficulty stemmed from the habitat in which the corals and brachiopods would have lived. Studies of modern representatives have shown that corals and brachiopods are most abundant in clear, warm, shallow, sea water with a relatively firm substrate. If the conditions change appreciably, larval stages of these animals settle elsewhere and the adults die out at the original location. Geological evidence would suggest that the corals and brachiopods of the past were little different in their requirements. Unfortunately, the Avon Gorge area was within a few tens of kilometres of a land mass throughout most of the Lower Carboniferous and the depositional conditions seem to have varied quite considerably both in the area itself and short distances away from it. Periodic changes in sea level also occurred when the sea bed became exposed and eroded so there are many non-sequences in the succession, although this was not fully appreciated at the time when Vaughan did his work.

Figure 3.2 shows some of the changes in lithology within the Avon Gorge section and also indicates some of the recently proved non-sequences in the succession. For comparison, the succession is also shown for the Forest of Dean, some 40 km to the north, (non-sequences not indicated). The thinning out of the deposits on nearing the old land mass of 'St. George's Land' is clearly evident.

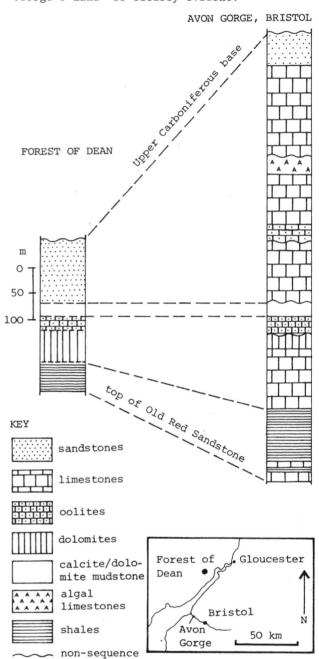

Fig. 3.2 Comparative vertical sections of the Lower Carboniferous of parts of south-west England.

A Modified System

Later work with corals and brachiopods has resulted in a modified system which overcomes some of the earlier deficiencies but still leaves unanswered the main problem of the effect on the faunas of changes in facies. The newer zones are shown in Table 3.1.

Figure 3.3 is a re-drawing of the Avon Gorge section relating the rock types to the modified zonal names.

How feasible is it to apply these modified zones to the rest of the country? The map (Fig.

TABLE 3.1 THE MODIFIED CORAL-BRACHIOPOD ZONES OF THE LOWER CARBONIFEROUS.

NAME OF ZONE	CODE LETTER OF ZONE
Dibunophyllum	D_2
	D_1
Seminula	S_2
Upper Caninia	$C_2 S_1$
Lower Caninia	C_1
(Transition)	γ
Zaphrentis	Z
Cleistopora	K

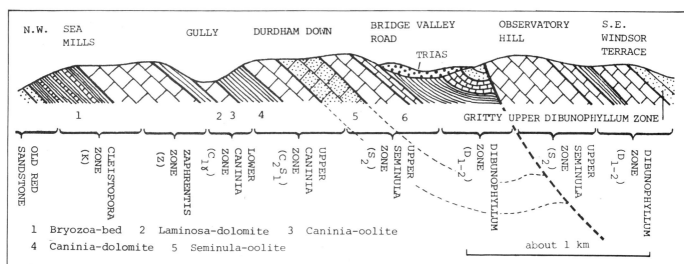

Fig. 3.3 *The Avon Section (right bank) showing Vaughan's zonal and lithological divisions.*

3.4) shows the diversity of environments which existed throughout much of the Lower Carboniferous in Britain. To the south of the Bristol area lay the marine 'Culm' basin of Devon and Cornwall, where shales and turbidites were accumulating. In such an environment most corals and 'shelf-type' brachiopods would be unable to survive, but goniatites are common enough to provide a valuable alternative. To the north, 'St. George's Land' formed a

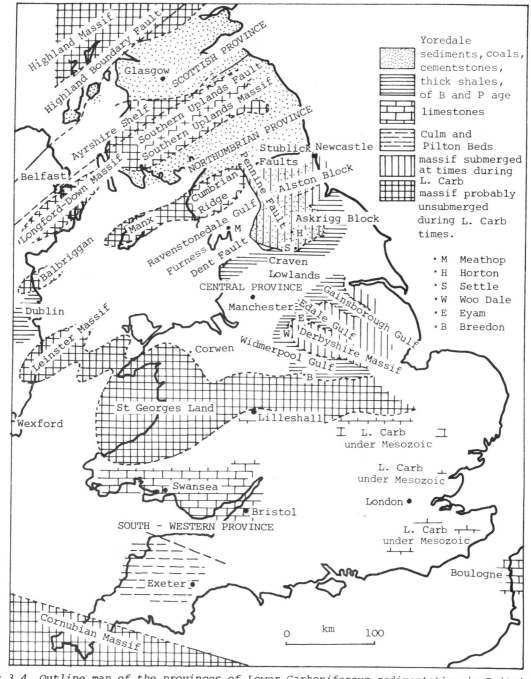

Fig. 3.4 *Outline map of the provinces of Lower Carboniferous sedimentation in Britain.*

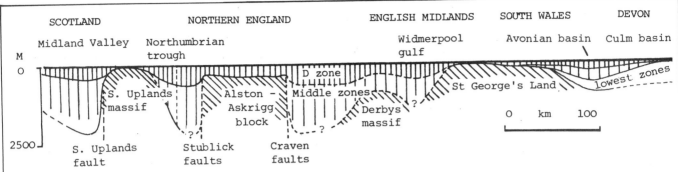

Fig. 3.5 Reconstructed and highly generalised section of the basins of Lower Carboniferous sedimentation in Britain. The maximum thickness is not known in the heart of the major troughs, or the continuity of sequence from possible underlying Old Red Sandstone. The line of section runs approximately southwards from Scotland to Derbyshire and thence southwestwards to Devon.

substantial barrier between the Bristol area and further shelf seas of the Central Province. Migration of faunas would therefore have been around the eastern and western ends of St. George's Land.

Within the Central Province itself there were considerable facies contrasts. Some areas were underlain by rigid 'blocks' or 'massifs' which seem to have subsided slowly as carbonate sediments were deposited on them, providing good habitats for corals and brachiopods. In between the blocks lay local 'troughs', or 'gulfs' where deposition was more rapid and impure limestones and shales were laid down containing goniatite and bivalve faunas. These goniatites and bivalves are not easily correlated with the corals and brachiopods of the 'blocks'.

Northwards again lay the Northumbrian and Scottish Provinces, which were sufficiently close to the old Caledonian land mass to receive a great deal of clastic sediment. Corals and brachiopods would therefore be expected to be few and far between in this area.

Figure 3.5 is a highly generalised section of the country showing some of these variations in Lower Carboniferous sedimentation.

A further complicating factor stems from the relative geographical isolation of the various areas of deposition. Evolution of organisms in such separated areas seems to have quite frequently proceeded at different rates. This makes it difficult to compare the faunas of different provinces for stratigraphic purposes.

In spite of these difficulties inherent in the modified Vaughan system, considerable advances in stratigraphic knowledge have been made through its use. Thus, it can be shown that the oldest parts of the succession (K and Z zones) are best developed in the South West Province. These older zones appear to be mostly absent from the 'blocks' of the Central Province although some strata of this age are suspected to occur in some of the troughs. On the 'blocks' therefore, the Lower Carboniferous is unconformable on a variety of older rocks and the zonal method has been used to try to date the local base of the Carboniferous System. Thus, in the Eyam borehole which penetrates to the hidden 'basement' of the Derbyshire block, the oldest Carboniferous strata are of C_1 age and rest on Ordovician rocks. At Thornton Force, near Ingleton in North Yorkshire, the superbly exposed unconformity consists of a basal boulder bed and limestone of S_2 age resting on vertically folded slates of possible Pre-Cambrian date (Fig. 3.6).

Fig. 3.6 Thornton Force, Ingleton.

A New Approach

The application of the coral-brachiopod zones is clearly restricted to relatively limited sequences of rocks. Are there any other techniques which might be applied?

In recent years much work has been done on microfossils, and spores have proved very useful for dating non-marine strata. In marine rocks valuable zonal systems have been set up using foraminifera and conodonts (tooth-like objects of unknown origin) but again there are difficulties in relating one set of zones to the other.

The most recent attempt at correlating rocks of the Lower Carboniferous has been derived from sedimentological research and is quite independent of the presence of particular zonal fossils. W.H.C. Ramsbottom and other workers have identified a number of 'cycles' of sedimentation which represent successive transgressions and regressions of the sea. These are of such widespread occurrence that they are generally believed to represent eustatic changes of sea level, i.e. probably world-wide changes, rather than local ones brought about by limited tectonic events. If

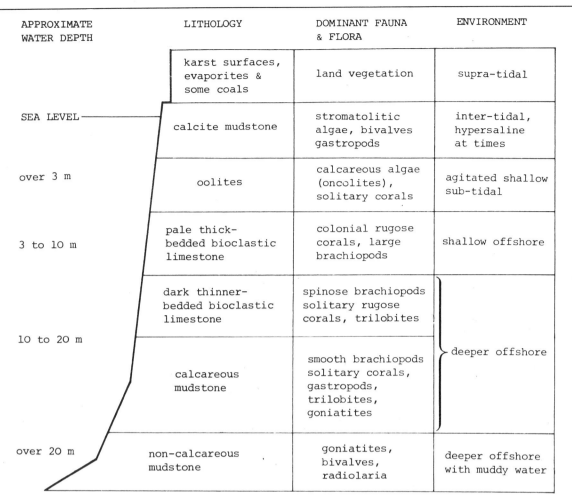

APPROXIMATE WATER DEPTH	LITHOLOGY	DOMINANT FAUNA & FLORA	ENVIRONMENT
	karst surfaces, evaporites & some coals	land vegetation	supra-tidal
SEA LEVEL	calcite mudstone	stromatolitic algae, bivalves gastropods	inter-tidal, hypersaline at times
over 3 m	oolites	calcareous algae (oncolites), solitary corals	agitated shallow sub-tidal
3 to 10 m	pale thick-bedded bioclastic limestone	colonial rugose corals, large brachiopods	shallow offshore
10 to 20 m	dark thinner-bedded bioclastic limestone	spinose brachiopods solitary rugose corals, trilobites	deeper offshore
	calcareous mudstone	smooth brachiopods solitary corals, gastropods, trilobites, goniatites	
over 20 m	non-calcareous mudstone	goniatites, bivalves, radiolaria	deeper offshore with muddy water

Fig. 3.7 Generalised section (not to scale) to show the changes in rock types and fossil content which indicate deepening or shallowing of waters in the Lower Carboniferous.

this is so they should provide universal 'marker' events.

The nature of a typical cycle is represented diagrammatically in Fig. 3.7, which also suggests very tentative water depths. It can be seen that at the time of maximum marine transgression the water may have been deep enough and sufficiently muddy for mudstones to form, containing a largely pelagic fauna. At other times thick bioclastic limestones were produced containing a diverse benthonic fauna. At times when the sea level was lower, widespread shorelines developed with thin evaporite sequences being produced in places. It has even been possible to identify ancient 'karsts', or weathered limestone surfaces, caused by subaerial weathering of the rock

following complete withdrawal of marine conditions for a while from some areas.

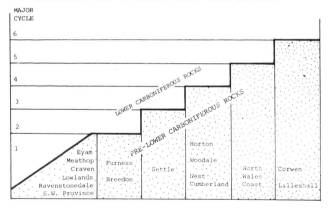

Fig. 3.8 Diagram illustrating the transgressive nature of each major cycle as the seas covered increasing areas of Britain in the Lower Carboniferous.

Six widespread major cycles of this type have been identified, as well as a number of minor ones. They are shown in Table 3.2, with very approximate equivalents under the Vaughan system.

One of the uses of the new system is illustrated in Fig. 3.8. This shows quite clearly how the Lower Carboniferous seas gradually encroached on more and more of the country, and also demonstrates that it was far from the simple south to north transgression envisaged by some earlier workers. (Map locations are shown in Fig. 3.4.)

TABLE 3.2 CYCLES IN THE LOWER CARBONIFEROUS.

CYCLE	APPROXIMATE EQUIVALENT ZONE
6th group of minor cycles	base of D_2 upwards
5th group of minor cycles	D_1
major cycle 4	much of S_2
major cycle 3	S_1 much of C_2
major cycle 2	lower parts of C_2 upper parts of C_1
major cycle 1	lower parts of C_1 Z K

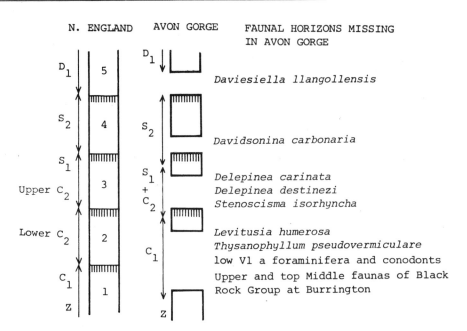

Fig. 3.9 Diagrammatic comparison between the Lower Carboniferous successions in northern England and the south west, showing the missing horizons in the Avon Gorge section. Numbers represent major cycles.

Rather ironically, the 'cycle' approach has revealed many gaps in the sequence at the Avon Gorge. Vaughan had picked this as his 'type section', believing it to be the most complete in the country, against which the rocks of other areas could be measured as a yard stick. Although this is probably true for the lower zones, the Avon Gorge bears poor comparison with northern England higher in the succession, as Fig. 3.9 shows. The presence of such un-suspected gaps in the record helps to explain why it was so difficult to apply Vaughan's zones universally and has led Ramsbottom to declare that, '... It would now seem that the Avon Gorge is one of the worst places to have taken as a type section ...' (1973). Of course, this statement was made with the wisdom of hindsight and the benefit of 70 years further study of the problem by many workers. Great advances have been made in our understanding of the evolution of corals and brachiopods, in sedimentological research and in the whole new field of micropalaeontology. Even today, however, we have to admit that we do not know all the answers and at the time of writing this Unit a furious battle is being waged through the pages of the scientific journals over the relationship between the 6 cycles and the 'stages' of the Lower Carboniferous established on the continent with the use of microfossils.

4

Methods of Zoning the Coal Measures

Introduction

COPY OF

PROPOSALS,

ENTERED INTO NOVEMBER, 24th. 1803.

FOR THE DISCOVERY OF

COALS:

At BREWHAM, *Somerset.*

FIRST. THAT the Sum of Two Thousand Pounds be rose, in One Hundred Shares, of Twenty Pounds each: Any Person may subscribe one or more Shares, half Share, or Quarter Share.

SECOND. THAT No Money be advanced till the One Hundred Shares are full, or the Two Thousand Pounds subscribed

THIRD. THAT as soon as the Subscription is full, Each Subscriber shall have Notice to attend a General Meeting of the Subscribers, for the purpose of appointing a Committee of Twelve of the Subscribers: No Person to be of the Committee that has not subscribed One Share, or Twenty Pounds.

FOURTH, THAT the Committee do appoint a Treasurer, and procure a Piece of Land, and Workmen, to make a Proper Trial for the Discovery of Coal.

\| Any Person may become a Subscriber, by application to JAMES JAMES, of COLE, near BRUTON.

MOTT, PRINTER, BRUTON.

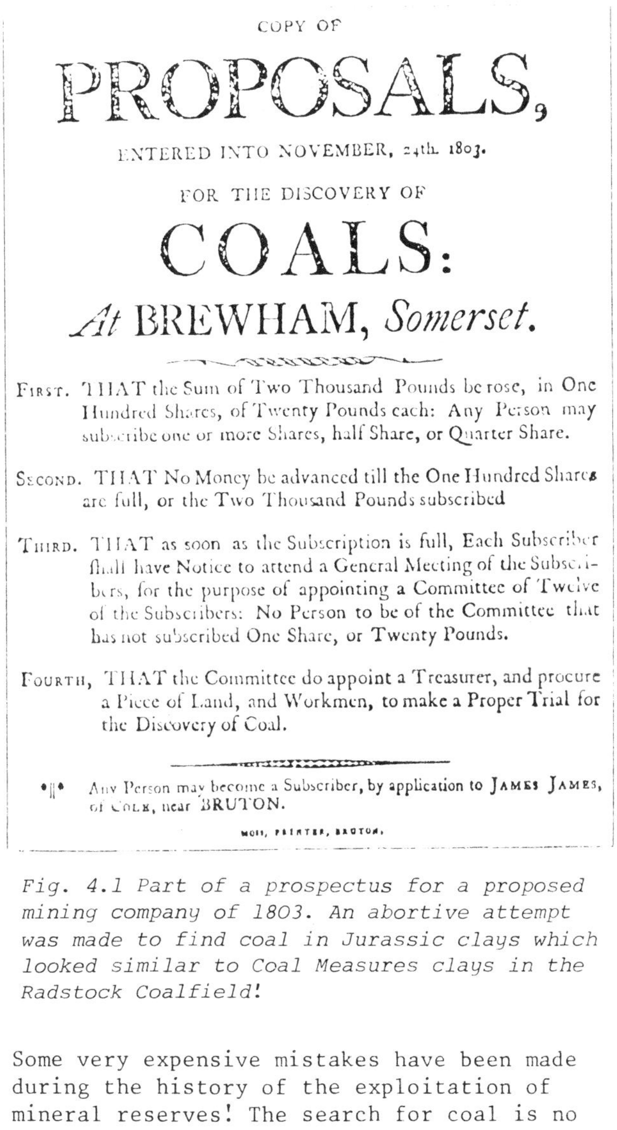

Fig. 4.1 Part of a prospectus for a proposed mining company of 1803. An abortive attempt was made to find coal in Jurassic clays which looked similar to Coal Measures clays in the Radstock Coalfield!

Some very expensive mistakes have been made during the history of the exploitation of mineral reserves! The search for coal is no exception. Fig. 4.1 is an extract from a prospectus issued by a company in the early nineteenth century hoping to sink a coal mine in Somerset. One can imagine the irate shareholders' meeting a few months later, when it was found that the miners had been digging through Jurassic clays, which simply looked like their Carboniferous counterparts, and all this investment had been wasted!

In contrast, the rich new coalfield of Selby in Yorkshire was completely missed at first, owing to poor borehole data obtained at the turn of the century. Now, however, the National Coal Board is spending over £550 million (at 1978 prices) to develop the field (Fig. 4.2). They must be very sure of their geological facts before committing that much money to the project! Our knowledge of stratigraphy and the use of methods of dating and correlation have advanced so much in recent decades that the authors make no excuse for including a second Carboniferous case study in this Unit. There are also new factors to be considered, since the Coal Measures are largely of continental origin, in contrast to the mainly marine successions considered so far.

Facts & Figures

- The Selby Coalfield covers 110 square miles — almost the size of the Isle of Wight
- Coal from the most distant workings will travel 24 kilometres (15 miles) and climb 1,000 metres (3,280 feet) — roughly equivalent in distance and climb to carrying it from Caernarvon Castle in North Wales to the top of Snowdon
- At full production a loaded railway train carrying 1,000 tonnes of coal will leave the Gascoigne Wood drift mine every half hour
- The 4,000 miners who will eventually work in the Selby mines will produce 10 million tonnes of coal a year. Productivity — at 12 tonnes for each manshift worked — will be four times the British average.
- Each of the five mines will have four coalfaces in operation simultaneously. The machinery and equipment on each face will weigh about 1,000 tonnes and will cost up to £3 million at 1978 prices
- Total cost of the scheme is £550 million (at 1978 prices)
- To prove the nature and quality of the coal seams at Selby some 60 boreholes were sunk from the surface. In addition seismic surveys — using sound bouncing techniques to plot the lie of the seams between the boreholes — were carried out on lines totalling 200 miles in length

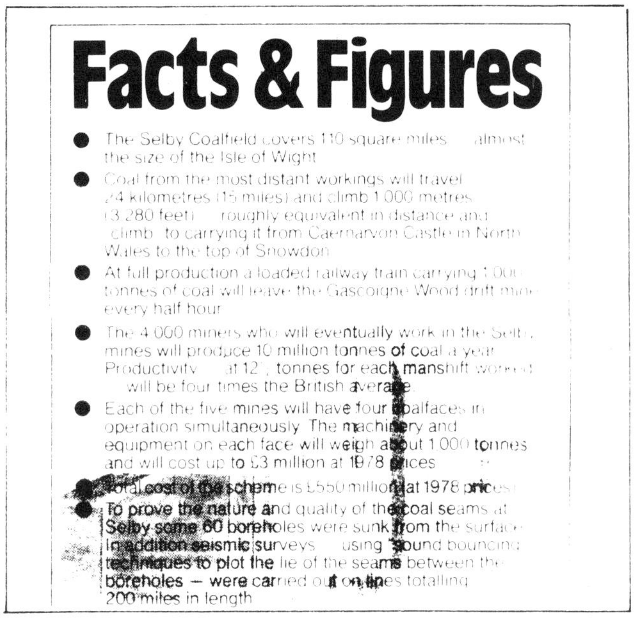

Fig. 4.2 Part of a publicity brochure about the Selby Coalfield issued by the National Coal Board in 1979.

Distribution of Coal Measures rocks

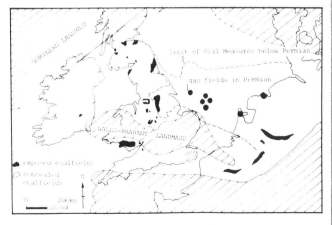

Fig. 4.3 Distribution of coalfields in north-west Europe, together with the area of the southern North Sea underlain by Coal Measures as proven by hydrocarbon exploration.

The map (Fig. 4.3) shows the main coal fields of Britain and northern Europe and distinguishes between 'exposed' coal fields where

Coal Measures strata crop out, and 'concealed' ones where they are buried beneath younger rocks. The location of natural gas fields in Permian beds is included, because the gas is thought by some to have originated in Coal Measures rocks below. The map also indicates the main land masses of Coal Measures times, in generalised form. In between these upstanding masses lay the main basins of deposition, predominantly marine in Devon but mostly of freshwater conditions in the rest of the region. Occasionally, however, the sea encroached right across these areas, depositing thin beds with quite distinctive marine fossils.

The coalfields of today are widely separated, but it seems most likely that the actual deposition of sediments was not limited to the present sites, but was spread across a more extensive area. These once widespread deposits were then folded and the crests of anticlines were eroded off, leaving the Coal Measures as remnants on the flanks of anticlines and in the synclinal basins. Such separation of the outcrops makes correlation more difficult, but it is still necessary, for the economic reasons stated before. How then is it done?

Methods of Zoning and Correlating the Coal Measures

TABLE 4.1 CLASSIFICATION AND ZONAL SCHEMES APPLICABLE TO THE BRITISH COAL MEASURES.
'West' is short for Westphalian, the continental stage name for most of the Coal Measures succession.

SCOTLAND	ENGLAND & WALES	MARINE MARKER BANDS		PLANTS		MIOSPORES	NON-MARINE LAMELLIBRANCHS
Macgregor 1960	Stubblefield & Trotter 1957	British Midlands	West Germany	Heerlen 1927,1935	Zones Dix 1934	Zones Smith and Butterworth 1967	Zones Trueman and Weir 1946
Upper (Barren) Coal Measures	Upper Coal Measures			West D	I	T. obscura	Prolifera
					H		Tenuis
— M —	— M —	Top (A. cambriense)		West C	G	T. securis	Phillipsii
	Middle Coal Measures	Mansfield Aegir (A.aegiranum) (A.hindi)			F	V.magna	Upper Similis- Pulchra
— M — Middle Coal Measures				West B	E	D. bireticulatus	Lower Similis- Pulchra
— M —	— M —	Clay Cross Katharina (A.vanderbeckei)			D	S.rara	Modiolaris
Lower Coal Measures	Lower Coal Measures			West A	C	R.aligerens	Communis
						T.sinani C.saturni	Lenisulcata
— — — — — —	— M —	Pot Clay Sarnsbank (G.subcrenatum)					

Table 4.1 summarises the main approaches to the zoning of the Coal Measures in Great Britain. The variety of methods is a clear indication of the care and attention which have been lavished upon these rocks, and the chart appears totally bewildering. We hope, however, to show how each method has been derived and the contexts in which each is of greatest value.

Lithological methods, 'the miner's way'

We have already seen how misleading and costly it can be to rely on rock types alone without reference to the fossils contained in them, but there are some circumstances where lithological characters are useful. This applies particularly to the nature of the coal itself, which varies from seam to seam in any one coalfield. At a school where one of the authors worked, the caretaker, who had formerly been a miner, provided an unrehearsed demonstration of this method. Pupils had each been asked to bring a piece of house coal from home to the lesson. Each one lived in a different mining village where the coal came from the nearest colliery, so there was considerable variation among the coal specimens. The caretaker happened to come in, his curiosity was aroused and he went around the room, identifying the seam from which each piece of coal had come, and in most cases naming the colliery! He was proved right with the majority of specimens and thus began an annual 'party-trick'!

Such techniques are only of local validity and cannot readily be extended to the other sediments of the Coal Measures, because of the repetitive cyclical nature of the deposition and the frequent facies changes. How, for example, could one set about correlating, on lithological grounds, the succession of the Yorkshire, Nottinghamshire, Derbyshire Coalfield with that of South Wales? Representative columns are given in Fig. 4.4. There is clearly very little comparison. This may seem an academic exercise, but what if a concealed coalfield were to be discovered somewhere in between? Which of the exposed fields would it more closely resemble and where should one look for the best coals? Just such an event has recently happened with the discovery of coal reserves beneath Oxfordshire. To some

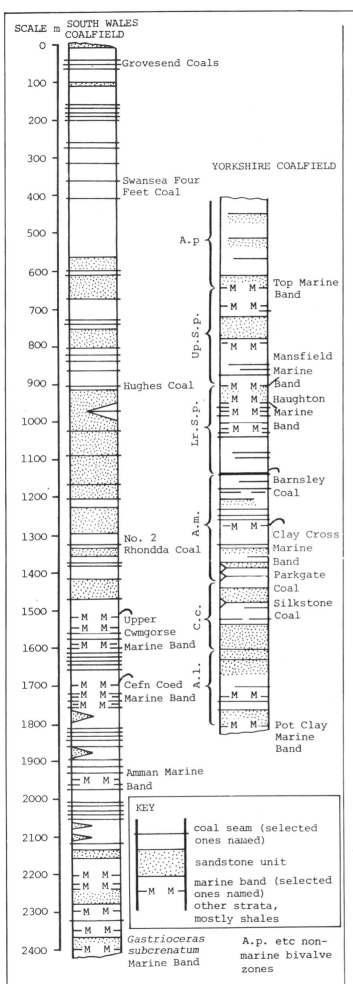

Fig. 4.4 Generalised sections of the Coal Measures of the South Wales and Yorkshire Coalfields showing the chief coal seams, sandstones and marine bands.

extent, the classification by the former Geological Survey into Lower, Middle and Upper Coal Measures was based on broad lithological differences, but changes in fossils have also been used to help in defining the actual boundaries.

The use of non-marine bivalves ('mussel bands')

At certain horizons, the Coal Measures abound with the fossils of fresh water bivalves and other animals. These tend to be concentrated in well-defined layers, known to the miners as 'mussel bands'. Until the 1940s, these were little used because it seemed to be too difficult to identify the fossils well enough to be sure of the species. Then, however, much laborious statistical work was done, using large numbers of specimens. This showed that if one allowed for considerable variation from the 'norm' for each species it was possible to distinguish one species from another. The stratigraphical ranges and relative abundance of some common mussel genera are shown by the black shapes in Fig. 4.5. The names by which the zones are known are given near the top of the diagrams. Their complete names are as follows:

Anthraconaia prolifera
Anthraconauta tenuis
Anthraconauta phillipsi   (A.p)
Anthraconaia pulchra   (Up.s.p)
Anthracosia similis   (Lr.s.p)
Anthraconaia modiolaris   (A.m)
Carbonicola communis   (C.c)
Anthraconaia lenisulcata   (A.l)

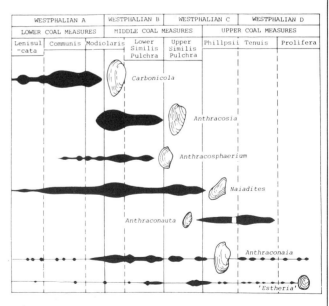

Fig. 4.5 Stratigraphic ranges and relative abundance of some non-marine bivalves and 'Estheria' in the Coal Measures.

Although, as usual, the name of only one species has been applied to each zone, in practice an assemblage of species is used. The method suffers from the disadvantage that large numbers of individuals are necessary, so it is not easily applied to borehole samples. There are also considerable variations in bivalve faunas from one coalfield to another.

The use of plant fossils

Of course, much of the land-plant cover of Coal Measures times is preserved as coal, the plants having been transformed first into peat and later into coal. These processes have rendered all but the spores of the vegetation virtually unrecognisable, but the shales which lie between the coal seams frequently contain well-preserved plant fossils. These comprise pith-casts of trunks and branches, impressions of roots or bark, imprints of a variety of leaves and various other structures such as seeds and cones. Figure 4.6 shows a selection of such plant fossils.

One of the problems of using plant fossils is that it is usually impossible to relate the various components together. Unlike the botanist's single name for one species of tree, whether he is looking at the bark or the leaves, the palaeobotanist may assign fossil impressions of bark or leaves to different species!

Plant fossils are also rather unevenly distributed, being more common in the coal-fields of Kent and the south-west than in the north. During their lifetime, there were notable facies contrasts too. For example, the club-moss *Lepidodendron* was most common growing in the clays of advancing lake deltas whilst the seed-fern *Neuropteris* was more characteristic of flood-plain sediments. Nevertheless, the palaeobotanists are able to set up plant zones which are of reliable use in the Coal Measures, although as Table 4.1 shows there are two 'rival' systems in use, depending on the criteria chosen!

The use of spores

Spores are the reproductive organs of plants, so they could perhaps have been included in the last section. However, they are of a different order of size, many of them being less than 200 microns across, so they require a range of special techniques for their extraction and the use of high-powered microscopes for their identification. They can seldom be directly related to the parent plant (except from cones) but this does not really matter, since they have an evolutionary history of their own which is reflected in their morphology. In common with the macro-fossils, evolutionary change is a 'one-way' process, so it is equally valid to anticipate the use of spores for relative dating.

It is only to be expected that the prolific vegetation of Carboniferous times should have produced countless spores, and that they should have been widely distributed by wind and water across the swamps and rivers and even into the sea. Fortunately they are extremely resistant to natural processes of decomposition and to the rather violent chemical separation techniques applied to extract them from the rocks. (See the Appendix of the Unit <u>Palaeoecology</u> for details.) Being so small, they can also withstand the grinding action of a drilling bit, so it is little wonder that they now form a formidable weapon in the coal geologist's armoury.

Table 4.1 shows the main zones based on miospores. Miospores (which probably had a male function) are smaller and less easily extracted than the female megaspores, but they exhibit more noticeable evolutionary change and are therefore used in preference.

Figure 4.7 shows the stratigraphic ranges of some common Coal Measures miospores. Again, it is normal to use concurrent ranges of spores, rather than seek only those which give their names to the zones. This is especially important in borehole logging where fragments of younger rocks are continually dropping down the borehole and 'contaminating' the drilling

Fig. 4.6 Typical fossil plants of the Coal Measures: 1) Part of the stem of a clubmoss, Lepidodendron x 1/4. *2) Part of the frond of a seed-fern,* Mariopteris x 1/2. *3) Leaves of a horsetail,* Annularia x 1/2. *4) Part of the frond of a tree-fern,* Asterotheca x 3/8. *5) Part of a fern-like leaf* Sphenopteris x 1/2. *6) Pith-cast of the stem of a horsetail,* Calamites x 1/3. *7) Part of the frond of a seed-fern,* Neuropteris x 1/2.

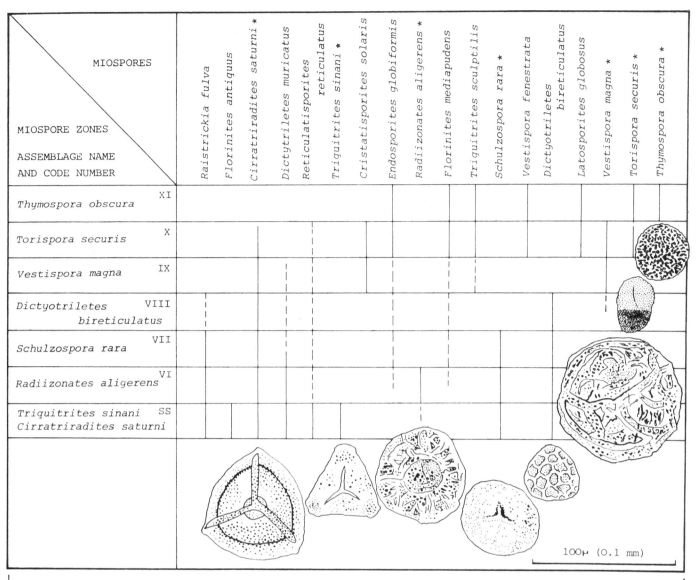

Fig. 4.7 The ranges of some Coal Measures miospores. Spores which give their names to the zonal assemblages are indicated by * and are illustrated. The scale is the same for all the spores.

mud which is conveying the chippings back to the surface for examination! Not only are large numbers of individuals present, but each assemblage is characterised by many different species.

The use of marine fossils

Much of the preceding section has been devoted to overcoming the difficulties of zoning and correlating the very variable suites of rocks deposited in non-marine conditions. From time to time, however, widespread changes in sea-level took place, probably of a eustatic nature and the sea covered extensive areas of low-lying country. Such marine incursions brought with them associated faunas, notably goniatites, marine bivalves and brachiopods (Fig. 4.8). Presumably, these animals had been evolving steadily in the distant seas, so the evolutionary record preserved in the marine bands of the coalfields is a rather disjointed one. In a way, this is useful because it makes each band quite distinctive.

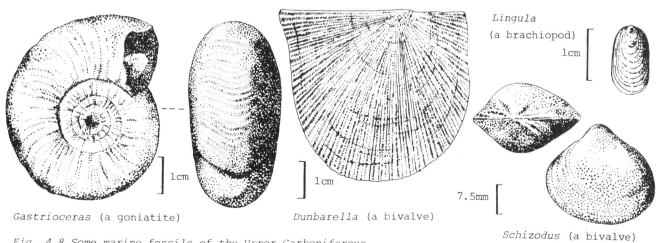

Fig. 4.8 Some marine fossils of the Upper Carboniferous.

TABLE 4.2 SOME IMPORTANT MARINE BANDS

SOUTH WALES COALFIELD	INDEX SPECIES OF GONIATITE	YORKS, DERBYS, NOTTS COALFIELD
Upper Cwmgorse	*Anthracoceras cambriense*	Top
Cefn Coed	*Anthracoceras hindi*	Mansfield
Amman	*Anthracoceras vanderbeckei*	Clay Cross
Gastrioceras subcrenatum	*Gastrioceras subcrenatum*	Pot Clay

Fig. 4.9 Main features of the faunal distribution in the Mansfield Marine Band in Britain. Coalfields are shown in black.

About 19 widespread marine horizons are known in the Coal Measures with others developed locally. Table 4.2 shows the 4 most important marine bands with the names of the most characteristic goniatite species. A host of local names has been applied to these important horizons: the ones shown in the tables are those in use in the Yorkshire, Derbyshire, Nottinghamshire coalfield and in West Germany.

Figure 4.9 shows the probable palaeogeography of the British area during the times of maximum marine advance at the level of the Mansfield Marine Band. In this map, the white areas represent the extent of the sea. Subtle facies contrasts existed within the sea, resulting in different assemblages of fossils from place to place, but fortunately goniatites are present in most areas. These are especially useful for correlation purposes, since they were nektonic in life and were therefore widespread and they also show evidence of rapid evolutionary changes.

We referred earlier (p.22) to the difficulty of correlating coalfields on the basis of lithological similarities alone. Turn back to the comparative columns of the South Wales and Yorkshire coalfields in Fig. 4.4. Any of the palaeontological methods of correlation between these two coalfields might be used, but you will probably find it simplest to use the marine bands. Draw a simplified copy of the columns marking, as a minimum, the position of the marine bands and, in generalised form, the levels where most of the coal seams occur. Using Table 4.2, draw lines between the columns to indicate equivalent times in the coalfields. What do you notice about the distribution of the main coal seams in each of the coalfields?

The Selby Coalfield

Many of Great Britain's older established coalfields are running out of economically obtainable coal seams. Coal still forms a major part of the energy base of Britain, however, so exploration for new reserves is still actively being carried out. (Project, Coal 2000.)

This brief case study is of the Selby Coalfield in Yorkshire, where development of a hitherto unsuspected coalfield is proceeding apace. It is intended that you should work through the study systematically and try to put yourself in the position of the geologists of the National Coal Board as the evidence began to emerge.

The Selby prospect was not at first a very promising one. Several boreholes had been drilled in the early years of this century to determine a possible north-eastern extension of the Yorkshire, Derbyshire, Nottinghamshire coalfield, but the records were disappointing. In the 1960s, however, another series of holes was drilled, which showed a much more encouraging picture, mainly because of improved drilling techniques, better recovery of cores

and 'lucky' siting. The siting of the holes was indeed virtually at random, since practically nothing was known of the sub-surface geology.

Figure 4.10 is a map of the area showing the location of the five holes drilled between 1964 and 1967. The boreholes penetrated varying thicknesses of Drift and rocks of Permo-Triassic age, before entering the Coal Measures at the depths below surface shown at the top of each column in Fig. 4.11. The entire Coal Measures succession down to the bottom of the holes is shown in Fig. 4.11. Selected coal seams are marked, mostly those 0.75 m or more in thickness. Clearly, many coals are present, some of them being in workable thicknesses, but it is not enough to merely <u>locate</u> them in each borehole. We need to know which seam is which, how each seam relates to those being worked in the nearest 'known' coalfield (in this case the Yorkshire field), and how the seams in each borehole may be correlated between the holes. Unless this is done, no inkling of the structural situation can be obtained.

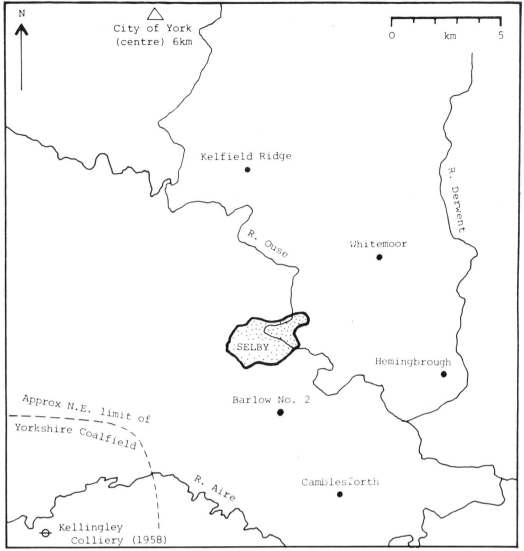

Fig. 4.10 Location of boreholes

TABLE 4.3 SIMPLIFIED SUMMARY OF BOREHOLE DATA.

Borehole	Kelfield Ridge	Whitemoor	Hemingbrough
Surface elevation in m above O.D.	8	7	6
Depth below surface of base of Permian (m)	388	497	566
Depth below surface of seam A (m)	728	867	806

At the time when the holes were drilled, the use of spores for zoning was in its infancy and the technique was not applied to the initial interpretation. The older methods were therefore used. The boreholes were cored throughout their passage through the Coal Measures and some good macro-fossils were obtained. Figure 4.11 shows the horizons of several widespread marine bands, whose names have been derived from localities elsewhere in the region, where they are particularly well developed. The assemblages of fossils in these marine bands in the Selby boreholes were distinctive enough for the geologist to be quite sure of the naming of the bands.

The horizons of the main bands of non-marine bivalves encountered in the boreholes are also shown. Alongside certain of these bands, particular bivalves have been indicated by numbers (see key) to enable you to carry out your own zoning, with reference to the range diagram for non-marine bivalves (Fig. 4.5).

EXERCISE

Now attempt the following using, where necessary, a tracing of the borehole logs:

(a) Draw lines between the boreholes to correlate the named marine bands.
(b) Estimate the stratigraphic horizon of the coal seam A, with reference to the marine bands and non-marine bivalve zones (Fig. 4.5 and Table 4.1).
(c) Determine the name of the seam by reference to its equivalent in the Yorkshire Coalfield (Fig. 4.4).
(d) What happens to the seam in the southern part of the area? (i.e. Barlow No. 2, Hemingbrough and Camblesforth boreholes).
(e) Read the following extract describing the extent of Seam A, from the original paper by R.F. Goossens (1973). 'The line of split (of Seam A into two thinner seams) has yet to be

proved firmly, but it is likely to run east-west through Selby. The western limit of the seam will be where it is terminated by the Base of the Permian. To the east the seam could well continue for some miles east of Whitemoor but of course will increase in depth to the point when working could become difficult. To the north the speculation still exists; a series of large faults could throw the seam out below the Permian, but it seems likely that it stretches for several miles, at least as far as the southern outskirts of the city of York.'

It would seem that with present technology, the maximum economic working depth is about 1100 m below the surface.

Use the above information and the simplified summary of the borehole data in Table 4.3 to estimate the working limits of the new coal-field, assuming that only Seam A is to be worked. You will be able to do this by three-point stratum contour (strike line) techniques on a tracing of the map (Fig. 4.10). You will need to draw stratum contours for the base of the Permian and for the Seam A. Use the data from Kelfield Ridge, Whitemore and Heming-brough only, and assume no faulting between them. Barlow No. 2 and Camblesforth are omitted from this exercise because large faults are suspected south of Selby, and as you have seen, Seam A splits in the southern part of the area. Calculate the average dip of Seam A from your stratum contours. When you have finished, check your answer in the Appendix and Fig. A.2.

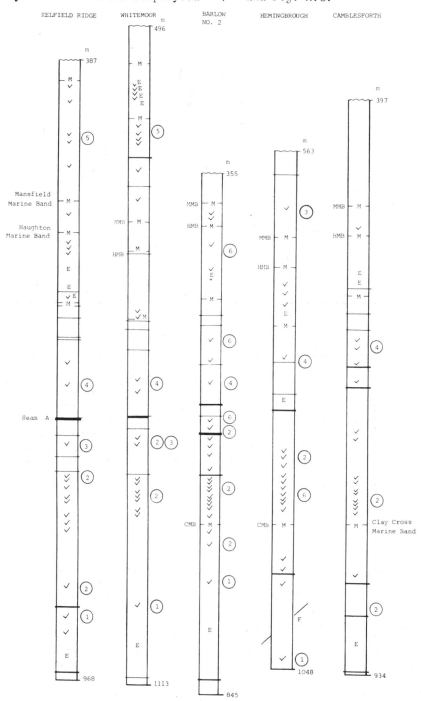

Fig. 4.11 Coal Measures successions recorded in boreholes of the Selby area.

M marine band; some named in full or by initial letters ✓ non-marine bivalve band E 'Estheria' band

——— coal seam of more than 0.75 m thickness ① etc. non-marine bivalves as follows:

1 Carbonicola communis 2 Anthraconaia modiolaris 3 species of Naiadites 4 Anthracosia similis
5 Anthraconaia pulchra 6 species of Anthracospherium F fault crossing borehole

5

Microfossils and North Sea Oil

Introduction

In previous sections of this Unit we have seen examples of occasions where a variety of fossils have proved useful in dating the rocks. Most of the fossils have been large fossils, the so called macrofossils, such as trilobites, graptolites and bivalves. However, more recently much smaller fossils, such as spores and pollen, collectively known as micro-fossils, have been found to be useful for stratigraphic purposes.

It is not difficult to understand why, traditionally, macrofossils have been the dominant type of fossils studied. Little was initially known about microfossils because their size required the use of a microscope whereas macrofossils can be seen, collected and studied relatively easily. Equally, it is not difficult to appreciate why microfossils have become exceedingly important. There are few sediments which do not contain microfossils of one type or another and instead of search-ing through the rocks for specimens as is often necessary with macrofossils the micro-palaeontologist can simply collect a two kilogram sample of sediment and take it back to the laboratory where it can be treated in a variety of ways to reveal its microfossil content under the microscope.

Micropalaeontological studies over the last two or three decades have received a tremen-dous boost as in one particular field they have proved invaluable. When drilling a bore hole, such as in the search for oil and gas, small chippings of rock are brought to the surface in the drilling mud. There is only a very remote chance that these chippings will contain any macrofossils or at least ident-ifiable fragments of macrofossils, but hun-dreds of complete and identifiable micro-fossils which can be used for dating will be contained within the chippings.

A North Sea oil well can cost more than £100,000 a day to operate and a complete well may cost over £3 million. It is therefore vital that the oil men know exactly where they are in the succession and do not waste time drill-ing beyond the depth they need.

Geophysical surveys help to give the oil men a very good idea of the structure they are about to drill into, but geophysical surveys do not reveal the age of the rocks and they can be open to different interpretations. (Examples of geophysical methods and their interpretations are given in the Unit <u>Geophy-sics</u>). Consequently a constant check on the age of the strata being drilled through is carried out using microfossils.

The extraction of microfossils from the rock

samples can be a time-consuming business and confirmation of the age of the rocks may be delayed. Generally this does not really matter as by noting the lithology of the rocks, taking account of geophysical surveys, and correlation with other wells, the oil company will know the approximate age of the rocks through which they are drilling. However, there are occasions where speed in getting the date of a rock is vital. One oil company was drilling recently through a succession of Mesozoic sediments which lay above Carbonifer-ous basement rocks. When they hit a coal meas-ures type cyclothem they wanted to know if these rocks belonged to the Upper Carbonif-erous or the very similar rocks found in the Mid Jurassic. If they were Mid Jurassic it would be worthwhile continuing drilling as the reservoirs in the area had been in Mid or Lower Jurassic sediments. However, if the sequence was Carboniferous there would be no point in continuing drilling as they would be in basement rocks below the potential reservoirs.

In order to get a rapid answer to their problem a sample of the shale from the sequence was taken by helicopter to Aberdeen and then by taxi to Sheffield University where some micropalaeontologists were wait-ing to analyse the sample. In extreme cases shale can be broken down in two hours using boiling hydrofluoric acid, although the process usually takes considerably longer than this. The spores obtained from this shale were analysed and it was a matter of a few moments to say whether these were Jurassic or Carboniferous (See Fig. 5.1 and 5.2). The

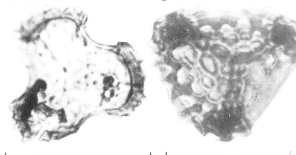

Fig. 5.1 Tripartites vetustus – *a typical Carboniferous spore.*
(Line = 50 μ)

Fig. 5.2 Ischyosporites variegatus - *a typical Jurassic spore.*
(Line = 50 μ)

spores showed that in this particular case the shale belonged to a Jurassic sequence and so the rig was alerted and advised that it was worthwhile continuing to drill.

Cases like this are unusual but it illustrates the important role that microfossils play in dating the rocks.

Microfossils Useful in Dating

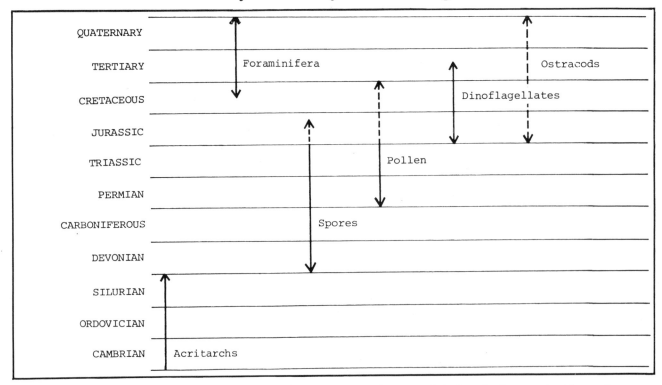

Fig. 5.3 Times when various microfossil groups are most useful for dating. N.B. These do not coincide with the total time ranges of the various groups.

As with the macrofossils, different types of microfossils are important at different levels within the stratigraphic record. Figure 5.3 gives some of the main microfossils that have proved useful for each particular Period in the North Sea. The list by no means includes all the microfossils which have stratigraphic potential.

Some of the names of these microfossil groups may be unfamiliar and so a brief outline of the various groups is given below.

Acritarchs are a varied collection of hollow, organic, walled, unicellular bodies 20-150 microns in diameter and of unknown affinities. The wall may have an opening in it revealing a central cavity suggesting that it is a cyst of some organism. Fossil acritarchs appear to be limited to marine strata and they date back well into Precambrian times. They are useful

to stratigraphers in the Lower Palaeozoic where they reached their acme and the choice of other microfossils is limited. Recent acritarchs are found and these also include non-marine examples. Acritarchs resemble a number of different objects to which they may have affinites, such as egg cases of invertebrates, dinoflagellate cysts and spores. The fact that the precise affinites of the acritarchs are uncertain does not limit their stratigraphic uses (Fig. 5.4).

Dinoflagellates are single-celled marine organisms 20-150 microns long. They are generally considered to be plants as they contain cellulose, although they do have some animal characteristics, having methods of propulsion. The first dinoflagellates are from the Triassic but Palaeozoic acritarchs may represent an earlier stage in dinoflagellate history. Modern dinoflagellates have a planktonic motile stage and a cyst stage but it is only the cysts that

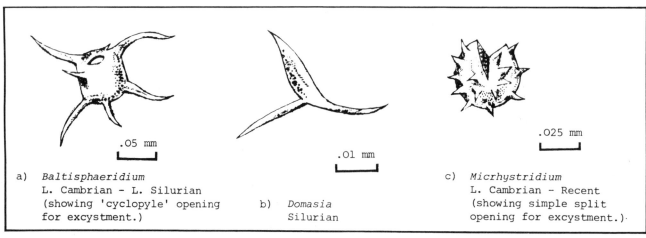

a) *Baltisphaeridium*
    L. Cambrian - L. Silurian
    (showing 'cyclopyle' opening
    for excystment.)

b) *Domasia*
    Silurian

c) *Micrhystridium*
    L. Cambrian - Recent
    (showing simple split
    opening for excystment.)

Fig. 5.4 Some Lower Palaeozoic Acritarchs.

are preserved as fossils (Fig. 5.5). These cysts are not found in many modern dinoflagellates but they seem to occur as a response to adverse environmental conditions such as low winter temperatures. The motile stage will emerge from the cyst in spring.

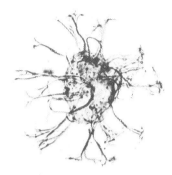

Fig. 5.5a Gonyaulacysta jurassica - *a typical Mid Jurassic dinoflagellate, showing archaeopyle through which motile stage escapes from cyst.*
Fig. 5.5b Surculosphaeridium vestitum - *a typical Upper Jurassic dinoflagellate*

Foraminifera are a very varied group of single-celled marine creatures with both planktonic and benthonic members. The soft body is enclosed in a test which may be either secreted or composed of agglutinated particles (Fig. 5.6). Foraminifera have existed from Cambrian or possibly even late Pre-Cambrian times but their usefulness for stratigraphic purposes has been mainly from the Upper Cretaceous onwards.

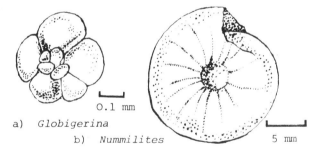

a) *Globigerina*
0.1 mm
b) *Nummilites*
5 mm

Fig. 5.6 Some Tertiary Foraminifera.

Ostracods are small crustacea with a shell formed of two valves (Fig. 5.7). They have become adapted to a wide variety of environments from marine through brackish to fresh water and even humid forest soils, but they are most commonly found in shallow marine areas. Ostracods are known from the Cambrian to the present but they are useful stratigraphically in rocks from the Jurassic to the Pleistocene. Species of Ostracods tend to have a restricted time range but also, unfortunately, they are often rather restricted in their distribution which limits their usefulness.

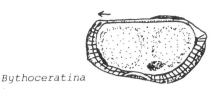

Bythoceratina

Fig. 5.7 A Mesozoic Ostracod.

Spores and Pollen are more familiar terms to most people but the distinction between them may not always be clear. A spore is a single-celled or few-celled body produced as a means of propogation in plants, bacteria, fungi and algae. However, most of the spores in the fossil record belong to vascular land plants as they possess a resistant cell wall. In some of the lower plants such as ferns the spores are of one type which gives a gametophyte with both male and female cells. Fertilisation takes place and the mature plant develops.

In large and more advanced plants such as *Lepidodendron* of the Carboniferous there are two types of spore, a male microspore 20-50 microns in diameter and a larger female megaspore 200-400 microns in diameter. These are dispersed by the wind and develop into independent gametophytes. Moisture is then required to carry the sperm to the female which then grows into the mature plant.

With higher seed-bearing and flowering plants the need to transport the sperm in moisture is eliminated, thus encouraging the plants to inhabit drier places. The female megaspores become trapped within an ovule and these are directly fertilised by modified multicellular microspores known as pollen grains which are transported to the female part of the plant by wind or water or insects (Fig. 5.8). The fertilised seed is then dispersed in its protective covering to develop into a new plant.

Fig. 5.8 Abietineaopollenites microalatus - *a typical Jurassic pollen. (Line = 50µ)*

Spores and pollen are designed to be dispersed over wide areas, being carried by wind or water or animals and this makes them important stratigraphically. The main periods in which they have proved useful are illustrated in Fig. 5.3.

Exploration in the Piper Oilfield

The Piper Oilfield provides just one example of where micropalaeontological studies have helped in the search for oil. This oilfield is relatively close to shore, lying about 150 km from the shore in the Moray Firth area (see Fig. 5.9).

Initial seismic surveys showed that the

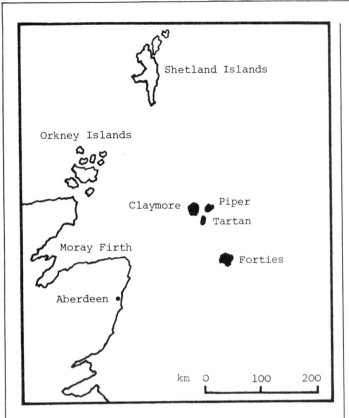

Fig. 5.9 *Location of the Piper Oil Field.*

general structure of the area was promising and so an oil company acquired six licence blocks in this sector of the North Sea in 1972. Three of these blocks contained large domal structures over a Carboniferous basement and these seemed the most likely oil traps. It was on these domes that the initial exploration was concentrated. The first well, from a drill ship, tested a large faulted anticline but no oil was found. In the second structure tested, a non-commercial quantity of oil was found in very thin sand of Jurassic or Lower Cretaceous age.

Originally the Company had hoped to locate oil in Palaeocene sands like the nearby Forties field but after the first two disappointments the next wildcat in the third anticlinal structure was primarily aimed at Mesozoic sandstones. This was because of the traces of oil in Mesozoic sandstones in the second well and also because seismic evidence had shown the Mesozoic section to be thicker over this area. Unfortunately the first hole into this structure from the drill ship had to be abandoned at 457 m because of anchoring and casing problems.

The next hole was made in November 1972, not by the drill ship but by a semi-submersible rig that had been moved into the area and which was more suited to the winter conditions. This drilled about 1 km from the original site and hit a highly porous and permeable oil sand of Upper Jurassic age between 2316 and 2379 m just before Christmas.

Production tests in the discovery well indicated that there were commercial quantities of oil. However, further appraisal wells would have to be drilled to establish the extent of the field. But before these were started the discovery hole (Fig. 5.10) was continued down into the Carboniferous basement. It was drilled this far down to establish the complete succession in the area and thus make it easier to establish the stratigraphic level and correlate the future wells. In all, seven further boreholes were drilled and most of these were terminated on reaching the Permian, although two were terminated in Mesozoic rocks below the Reservoir Sandstone. These boreholes confirmed the extent of the reservoir and showed that there was a sufficient quantity of oil to make it a commercially viable field. The lithologies encountered suggested that the reservoir sandstone was a beach or barrier bar complex 76 m thick on average and deposited above the wave base during a transgression onto Mid Jurassic non-marine sediments.

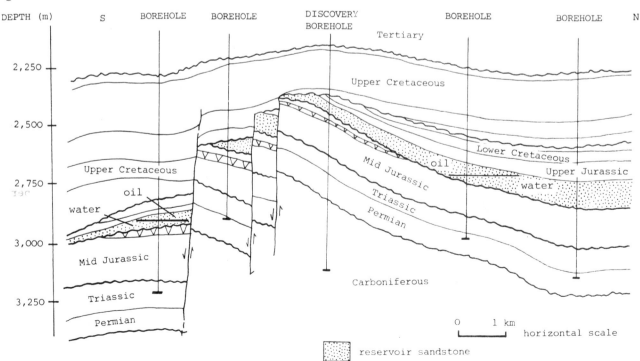

Fig. 5.10 *Section to show the structure of the Piper Oilfield and location of some of the boreholes.*

All this may seem quite straightforward, especially when looking at the cross sections. But do not forget that this information has been deduced at over 2,000 m distance. The original seismic surveys and later, more detailed surveys would have shown the general structure, but even this may have been open to more than one interpretation (see the Unit Geophysics where analysis of seismic data is dealt with). The actual lithologies and ages of the rocks can only be confirmed by drilling. This is where the microfossiis play their crucial role. Not only can the microfossils be used to date and correlate the rocks but many microfossils will give information about the environment of deposition as well. Once the succession has been established the microfossils are used to check the stratigraphic level which the borehole has reached, for if drilling were to continue below the potential reservoir rocks it would be a very expensive mistake. This is why there has to be a constant monitoring of the microfossil content of the chippings recovered from the borehole.

Appendix

Graptolites and Trilobites in the Ordovician

<u>Answer to the exercise on page 10</u>

Figure A.1 represents the standard zonal system for this part of the Ordovician. Sketches of each of the main zonal indices are given, although the usefulness of the other members of the assemblage in each zone must not be overlooked.
Didymograptus speciosus would be locally useful as a sub zone index. The zone of *Glyptograptus teretiusculus* is a concurrent range zone, the others are range zones, within the rather simplified limits of this version.

<u>Answer to the exercise on page 13</u>

(a) The most likely zone is that of *Nemagraptus gracilis*.
(b) There are no 'stray' specimens from outside this stratigraphic range, so the waste is probably all from the same quarry.
(c) The fossils are nearly all broken and jumbled together in a death assemblage. They show no preferred orientation (except for the two *Ogygiocarella* specimens, which are on different bedding planes and the orientation is probably fortuitious).
(d) The dark siltstone is suggestive of a low-energy environment of deposition. The mud-laden water would presumably have transmitted little light and it is probably significant that the majority of the trilobite specimens are of the blind trinucleid type. Of the sighted trilobites, many of the specimens are most likely the remains of moulted carapaces which could well have drifted into the area of deposition from a better-lit portion of the sea-floor. However, the largely random orientation indicates the absence of any appreciable bottom current.
(e) The 'tuning-fork' graptolite demonstrates that the rocks at Builth Wells belong to a lower horizon in the Ordovician, in this case the zone of *Didymograptus bifidus*. The regional dip of the strata is to the north-west, so one would expect the sequence to become younger from Builth Wells to Llandrindod Wells.

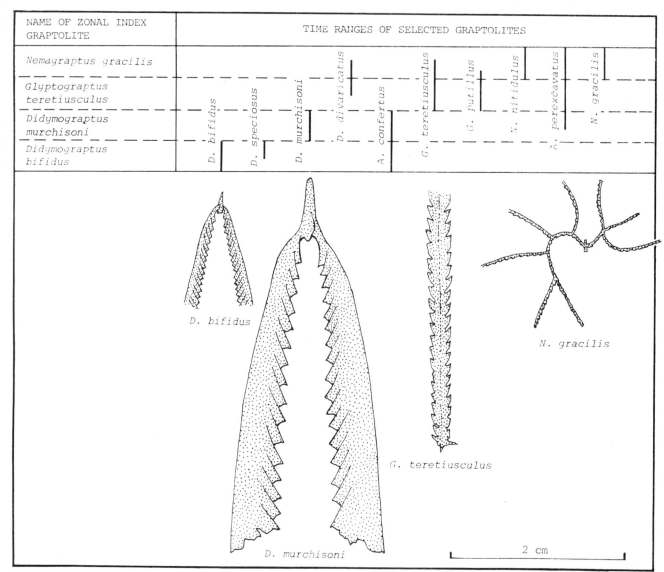

NAME OF ZONAL INDEX GRAPTOLITE	TIME RANGES OF SELECTED GRAPTOLITES
Nemagraptus gracilis	
Glyptograptus teretiusculus	
Didymograptus murchisoni	
Didymograptus bifidus	

Fig A.1 *The standard grapolite zones for part of the Ordovician.*

The Selby Coalfield

Answers to questions on page 28

(a) Correlation of marine bands

The Mansfield & Haughton Marine Bands can simply be joined between the boreholes by straight lines. The Clay Cross Marine Band was not recorded at Kelfield Ridge and Whitemoor, but the columns in Fig. 4.11 have been so arranged that the horizon of the Clay Cross Marine Band lies along the horizontal line continued from the other three holes.

(b) The horizon of seam A

The seam lies at the junction of the *modiolaris* zone and the lower *similis-pulchra* zone in the Middle Coal Measures.

(c) The name of the seam

Seam A is the Barnsley Seam. This was one of the Yorkshire coalfield's 'prize' seams, being thick, good quality coal with few impurities, but it has been largely worked

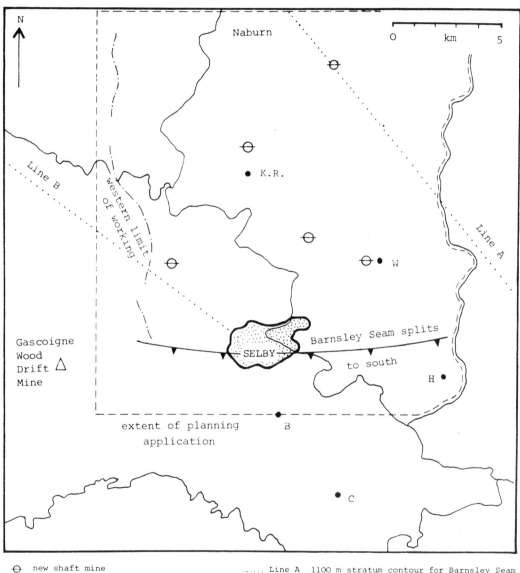

Fig. A.2   Sketch map of the limits of the Selby Coalfield.

out in the existing coalfield. It was also known to split into several less worthwhile seams in the north-eastern part of the coal-field (e.g. some 10 km to the south of Kellingley Colliery). The discovery of such a famous seam in prime condition in the Selby prospect was therefore particularly welcome, if rather surprising. Throughout most of the coalfield it is between 1.8 and 3.4 m thick.

(d) The Barnsley seam in the south of the area

The seam splits into two in the southern part of the Selby field. In Barlow No. 2 borehole both parts of the seam were found, but at Hemingbrough and Camblesforth the lower part has been 'washed out' by a channel, later filled with sandstone.

(e) The limits of the coalfield

The map (Fig. A.2) shows the limits of the area beneath which the National Coal Board has obtained permission to mine. It will probably differ in some respects from yours. (Lines A and B represent the limits which you should have obtained from your stratum contours). It was drawn up after a further 55 boreholes had been drilled and an extensive seismic exploration programme had been carried out in the early '70s. The latter is particularly useful for locating faults. The most northerly proving of the Barnsley seam was at a depth of 680 m at Naburn. Planning permission to work the Barnsley seam was submitted in 1974 and approval has not been sought to work the other seams, which are likely to be less economic. The first shafts were begun in 1977 and the first coal-production is planned for the early 1980s. It is estimated that the Barnsley Seam contains some 600 million tonnes of extractable coal and it is anticipated that output will reach 10 million tonnes per year.

The average dip of the Barnsley Seam, from your stratum contours, should be about 4°, or 1 in 13 to the north-east.

The new shafts shown on Fig. A.2 will be used to ventilate the workings and to provide access for the miners. The coal will all be brought to the surface at Gascoigne Wood Drift Mine. This is sited beyond the working limit of the coalfield but reaches the Barnsley Seam by means of a sloping tunnel, or drift.

Further Reading

The following sources may be useful to those who wish to read more widely:

Brasier, M.D.
Microfossils. Allen and Unwin, 1980. An advanced text covering all the main micro-fossil groups.

Clarkson, E.N.K.
Invertebrate Palaeontology and Evolution. Allen and Unwin, 1979. An advanced systematic text which summarises the stratigraphic value of each major fossil group at the end of each chapter.

Donovan, D.T.
Stratigraphy: An Introduction to Principles. Wiley, 1966. Contains a chapter on correlation by fossils, with several detailed examples.

Eicher, D.L.
Geologic Time. Prentice Hall, 1976. An introductory text covering all the main methods of establishing a geological time scale.

Kirkaldy, J.F.
Fossils in Colour. Blandford, 1975. An introductory text, based on systematic lines. Uses of fossils are discussed briefly after the description of each group.

Open University S23 Block 3
Palaeontology and Geological Time. A 'do-it-yourself' course which includes some interesting studies, notably more details of *Micraster*.

Raup, D.M. and Stanley S.M.
Principles of Palaeontology. Freeman 1978. An advanced text which covers many fascinating aspects of palaeontology.

Thackray, J.
The Age of the Earth. Geological Museum, H.M.S.O., 1980. A short but colourful text which deals with the stratigraphic uses of fossils and includes pictures of the common life forms of each era.

References

Elles, G.L. and Wood, E.M.R.
A Monograph of British Graptolites. The Palaeontographical Society, London 1916.

Goossens, R.F.
Coal Reserves in the Selby Area. Midlands Institute of Mining Engineers, paper 4462, 1973.

Lapworth, C.
'The Moffat Series', *Quarterly Journal of the Geological Society*, vol. 34 pp. 240-346, 1828.

Ramsbottom, W.H.C.
'Transgressions and Regressions in the Dinantian: a New Synthesis of British Dinantian Stratigraphy.*Proceedings of the Yorkshire Geological Society*, vol. 39, 1973.

Smith, W.
A memoir to the map and delineation of the strata of England and Wales with part of Scotland. London 1815.

Vaughan, A.
'The Palaeontological Sequences in the Carboniferous Limestone of the Bristol Area.' *Quarterly Journal of the Geological Society*, vol. 61, pp. 181-307, London 1905.

Geochronology

Contents

Acknowledgements

The authors are most grateful to the following for their help in various stages of the preparation of this Unit:
Dr F. Spode for his critical reading of the manuscript; Dr R.J. Firman for advice regarding sources of information on the granites of the Lake District; Dr D. Livesey for correcting our rusty physics! Mrs J. Kay and Mrs B.G. Ross for typing.

Any remaining errors are entirely the responsibility of the authors.

We are also grateful for permission to reproduce the following illustrations:
Fig. 2.5 and Fig. 5.2 The Institute of Geological Sciences; Fig. 3.3 and Fig. 5.8 Raup, D.M. and Stanley, S.M. *Principles of Palaeontology*, W.H. Freeman and Co.

The following figures have been based on illustrations from the sources indicated:
Fig. 2.1 from a poster of the Zion Natural History Association, Springdale, Utah; Fig. 3.1 Lyall, C. *Elements of Geology*, John Murray; Fig. 3.2 Kirkaldy, J.F. *Geological Time*, Oliver and Boyd; Fig. 4.5 The Open University, *Historical Geology*; Fig. 4.6 Sparks, B.W. and West, R.G. *The Ice Age in Britain*, Methuen; Fig. 5.1 Bott, M.H.P. *The Geology of the Lake District* The Yorkshire Geological Society; Fig. 5.3 Anderton, Bridges, Leader & Sellwood, *Dynamic Stratigraphy of the British Isles*; Allen and Unwin; Fig. 5.7 Glaessner, M. "Distribution and time range of Ediacara Fauna". *Bulletin of the Geological Society of America. Vol 82*; Fig. 5.9 Eicher D.L. *Geologic Time*, Prentice Hall; Fig. 5.11 H.M.S.O. *The Age of the Earth*.

1
Introduction

Over the years, as geological knowledge increased, it was found necessary, for the sake of convenience, to devise some methods for comparing the ages of the rocks. At first, this was only a matter of saying which rocks were older or younger than other rocks and then placing them in order. This produced the so-called 'Relative Time Scale'. A system was devised on a global scale whereby rocks could be compared to this relative time scale: sub-divisons were made, based mainly on fossil content and the result was the construction of the stratigraphic column with its familiar names of periods.

It was not necessary to contemplate just how long a period of time was represented by this relative time scale. In fact, very widely differing views, as we shall see, were held on the scale of geological time. However, man's curiosity is such that he sought to place dates and limits on this time scale. Various methods were sought ranging from biblical accounts and rates of sedimentation to the cooling rate of the earth. However, it was not until the discovery of radioactivity that meaningful values could be added to the relative time scale. These dates, in millions of years, have become known as 'absolute' dates.

Following a description of the methods by which the relative time scale became established, the ways of estimating 'absolute' time are discussed in this Unit. A number of examples are used to illustrate some of the problems and limitations involved with establishing absolute dates for rocks.

The case studies have been chosen to cover a variety of geological situations, ranging from some familiar British granites to the ancient sedimentary rocks of southern Africa which contain some of the earliest recorded life-forms.

2
Relative Dating

Superposition and Way Up

It may seem quite easy to us, observing undisturbed layers of sedimentary strata, to state which is the oldest and which is the youngest rock in the sequence. Obviously, the rocks first deposited will be at the bottom, with the youngest rocks lying at the top of the sequence. However, this fact has not always been appreciated because of earlier ideas about how the earth was created. It was first recognised by men like Nicolaus Steno (1638-86), a Dane who realised that there were well-defined layers in the rocks. Steno first published his findings in 1669, after observing strata around Florence in Italy. Although his work did not make a great impact on the scientific world, the concept of youngest upon oldest - the Law of Superposition - soon became established and it is now a fundamental principle of geology. Indeed, the Geological Time Scale (see page 9) with which we are all so familiar has been established using this concept.

An area like the Grand Canyon provides an ideal locality in which this law can be applied most simply. Although there are some ages not represented, virtually undisturbed horizontal layers of rocks lie on top of the Precambrian rocks seen in the bottom of the canyon. These layers reach up to the Permian by the lip of the canyon and continue up to the Tertiary at the top of some nearby peaks (Fig. 2.1). The sequence of rocks can thus be very simply followed.

However, very rarely is such an ideal opportunity to study the sequence of rocks presented to geologists. Much more commonly, several outcrops have to be fitted together to build up the overall picture. For example, when the beds in an area have been gently tilted, a traverse across that area will be cutting across successively younger beds and the net effect will be the same as the simple vertical section in the Grand Canyon, but without the continuous exposure. For example, on a journey from Holyhead, in Anglesey, to London, one would start on rocks of Precambrian age and end on Tertiary strata, passing over rocks of virtually every age in between. An early cross-section (Fig. 2.2) drawn by

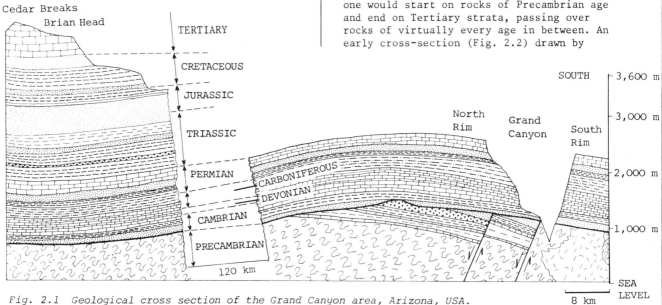

Fig. 2.1 *Geological cross section of the Grand Canyon area, Arizona, USA.*

Fig. 2.2 *Geological section from North Wales towards London. Vertical scale and gradients exaggerated*

William Smith in 1817 shows just this
succession, although it starts near Snowdon
rather than in Anglesey. The stratigraphic
column had not become established by this
time and so some of the names on Smith's
section may be unfamiliar, being mainly
lithological terms. However, by looking
at a modern geological map you should be
able to work out the ages of all these
rocks.

The Law of Superposition is simply applied in
areas where there are either horizontal or
gently tilted strata. But what of areas where
the rocks have been highly deformed? Correlating the various exposures to produce a
continuous section can be a difficult task and
at a particular exposure there is no guarantee
that the beds at the base of the exposure are,
in fact, the oldest beds. The whole section
might have been inverted. Older rocks can be
found above younger rocks in areas where there
has been thrust faulting or overfolding
(Fig. 2.3).

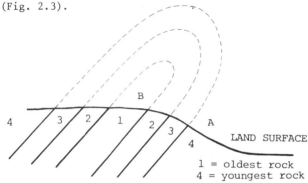

1 = oldest rock
4 = youngest rock

*Fig. 2.3 Overfold: illustrating that from A to
B the order of the rocks has been inverted, with
the youngest at the base of the sequence.*

Fortunately there are a number of ways in
which a geologist can check to see whether the
rocks have been inverted or not. The order of
occurrence of fossils may help but it is quite
likely that in a single exposure there would
be little variation in the fossil assemblages.
More usefully, there is a series of sedimentary structures collectively known as 'way-up'
structures which enable the original orientation of beds to be determined. Figure 2.4
illustrates just some of these 'way-up'
structures.

A. GRADED BEDDING

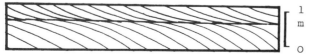

With a supply of mixed sediment the
coarsest material falls to the bottom first
and the finest material settles out on top.

B. CURRENT (CROSS) BEDDING

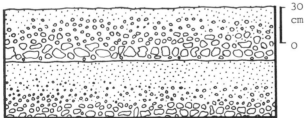

Material is deposited at an angle as an
underwater dune advances. In most examples,
the 'top set' of the current bedded unit is
truncated by erosion (as is shown in the upper
example) which makes it possible to distinguish
the top of the structure.

C. LOAD CAST

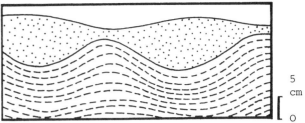

a) When a layer of sand is deposited over mud,
the weight of the sand may press down into
the mud and form a rounded depression.

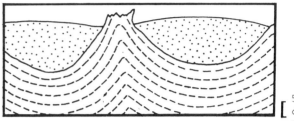

b) Sometimes the unequal loading may become
so great a 'flame' of mud breaks through
between the sand.

D. RAIN PRINT

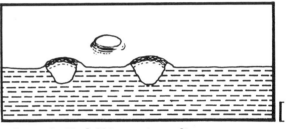

Rain or hail falling onto soft
sediment may leave a small crater.

E. MUD CRACKS

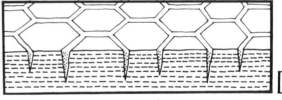

Also known as sun or desiccation cracks.
When wet, clayey sediment dries out, it
may contract and crack into a distinctive
hexagonal pattern.

F. WASHOUT

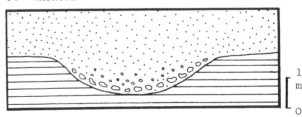

A Channel is cut into sediment and
filled with later material.

G. BURROWS

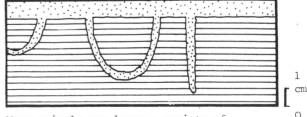

Many animals may leave a variety of
different-shaped burrows in sediment.

Fossils

Undoubtedly the most important method of establishing the order of the deposition of the rocks on a regional scale is not with superposition relationships, which can be difficult to follow over any distance, but by using fossils. The ways in which groups and species evolve, reach their acmes, decline and become extinct, are the most powerful tools at the disposal of the stratigrapher. Right from the early pioneering days of geology in the eighteenth and early nineteenth centuries, it was realised that fossils appeared and dis-appeared, in the sequence of rocks, in a fixed order and that this order was repeated in rocks of equivalent ages in many different areas. Rocks containing the same character-istic assemblage of fossils could therefore be correlated together, that is they could be said to be of the same age. A reasonably complete picture has now been built up of the time-distributions of the various fossil groups and individual species within these groups and so a fossiliferous rock can be dated relative to other fossiliferous rocks.

The use of fossils to date rocks is so important that another Unit has been devoted to the various techniques used to establish a stratigraphy from fossils (see the Unit Fossils and Time).

Unconformities and Non-Sequences

Further complications in determining the order of deposition arise when there have been breaks in the succession. Major unconformities are usually quite easy to recognise, as there is often some angular discordance between the two sets of rocks and two contrasting rock-types above and below the unconformity. In fact, the older set of rocks may be highly deformed with the younger set virtually un-altered (Fig. 2.5). Major unconformities can provide very useful reference points for dividing up the sequence of rocks. They are often caused by very significant orogenic events and therefore the same breaks may be found over a wide area. However, one problem is that it is usually very difficult to assess accurately just how long a period of time is represented by the unconformity and how many hundreds of metres of strata have been removed by erosion before the upper set of rocks was deposited.

Much more difficult to recognise in a succession of rocks is where there has been a break in the supply of sediment, but no deformation, so that there is no angular difference between the rocks. Many metres of sediment may have been removed, or not deposi-ted at all, during this time interval. These 'non-sequences', as they are called, can be detected in a number of ways. Detailed palaeontological surveys may reveal missing fossil zones, but more often they may be detected by structures along the bedding plane, such as wash-outs (Fig. 2.4), which would show that there had been some erosive activity.

Fig. 2.6 Borings into hard ground. Side view of oolitic limestone, Middle Jurassic, Ketton, Northants. The scale bar is 5 cm long.

Figure 2.6. illustrates another example of a structure that would indicate a non-sequence. These borings were made into 'hard-ground', that is, the rock was already lithified, showing that sediment must have been removed

Fig. 2.5 Unconformity: Triassic conglomerate rests unconformably on steeply dipping inverted Old Red Sandstone (River Avon).

from above before the animals bored into it. The bedding plane is coated with oysters so this was probably a wave-cut platform in the Jurassic period which, on exposure, was subjected to animal activity before being covered again by more sediment some unknown time later.

Recognition of a non-sequence shows that there is an incomplete rock record and there is a similar difficulty to that experienced with unconformities. Even after detailed comparisons with other sections covering the same period of time, it may be difficult to ascertain how long a time interval is not represented.

Cross-Cutting Features

Sometimes it is not practicable to determine the order of superposition and it is not always possible to use fossils to assign relative dates to rocks; for instance, there are normally no fossils in igneous rocks. Yet there are other methods that enable the rocks to be placed in order of age. Careful observations in the field may show relative age relationships. For instance, a dyke will obviously intrude and metamorphose only those rocks older than itself, it will have no effect upon rocks younger than itself (Fig. 2.7).

These 'cross-cutting' relationships can be seen not only in dykes but also in such features as mineral veins and faults, both of which will only cut rocks and structures older than themselves.

Fig. 2.7 Cross-cutting relationships. You should should be able to place the four rock-types illustrated in their correct chronological order.

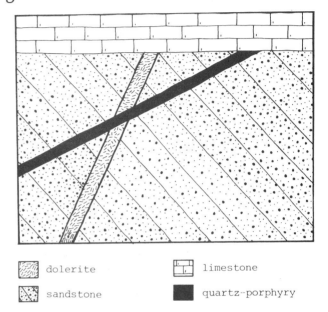

| | dolerite | | limestone |
| | sandstone | | quartz-porphyry |

Included Fragments

A conglomerate, or breccia, made up of rock fragments, which can be identified, also enables rocks to be dated relative to one another. Obviously the conglomerate must have been formed later than all the rocks that are represented by fragments in it (Fig. 2.8).

The same reasoning applies to xenoliths in igneous rocks. The igneous rock must be younger than the xenolith it contains. If the parent rock from which these were formed can be ascertained a relative date can be established.

Fig. 2.8 Triassic breccia containing fragments of Precambrian rocks, (Charnwood Forest, Leicestershire) and formed in a wadi.

An Illustration – The Shap Granite

The techniques outlined above are basically matters of observation and simple logic. They can be applied in many different situations, especially in the field, to establish the relative ages of the rocks.

The outcrop of Shap Granite on the eastern edge of the Lake District provides an excellent example of where a number of these techniques can be applied to give the 'relative' date of the granite. Later on in this Unit (p. 23) we shall demonstrate how absolute dating methods have been used on this same granite.

The outcrop is some 8 km^2 in area and the

granite has a distinctive appearance. It contains large pink orthoclase feldspar phenocrysts up to 5 cm in length. The granite is worked today, mainly for roadstone, although some is used for a variety of ornamental purposes. Fragments of the granite are well known throughout the north of England, often being found as glacial erratics.

The granite intrudes into and thermally metamorphoses rocks varying in age from Lower Ordovician to late Silurian. Many metamorphic effects are seen in the surrounding country rocks, hornfelses being produced as well as many new minerals developing within the aureoles as a result of metamorphism and meta-somatism. Thus, by simple 'cross cutting' relationships, the granite must be upper Silurian or younger (Fig. 2.9).

Xenoliths of both Ordovician and Silurian rocks, but notably the Ordovician Borrowdale Volcanics, have been recovered from the intrusion which, by 'included fragments', further confirms that the intrusion post-dates these rocks.

This established a maximum age for the granite, so what is now needed is some method of finding its minimum age. About 1.5 km due east of the intrusion, there is an exposure of conglomerate resting on Silurian rocks. This conglomerate is at the local base of the Carboniferous succession. Contained within the conglomerate are pebbles of pink orthoclase feldspar. These large feldspars cannot have been transported far, otherwise they would have broken down; feldspar is particularly susceptible to physical and chemical weathering. The only possible local source for the feldspars is the Shap Granite, which means that the intrusion of granite had become exposed and subjected to erosion by Lower Carboniferous times.

We now have our minimum age for the granite and are thus able to 'bracket' the date. It must have been intruded sometime between the Upper Silurian and the Lower Carboniferous. More detailed observations of the relationship of the granite to the cleavage in the country rocks enable this 'bracket' for the dates to become more precise. From numerous studies in and around the Lake District it is known that the main tectonic events of the Caledonian Orogeny took place during the late Silurian and early Devonian and this is when the main cleavage was formed. As the main cleavage has been deflected by the forceful intrusion of the granite, it follows that the granite must be later than the main deformation event. However, outside the aureole it has been observed that the main cleavage has been re-folded into kink-bands while these are not seen within the hornfelses of the aureole. It has been argued that the kinking is not seen simply because the hornfels was resistant to it. This means that the granite was intruded after the main phase of deformation but before the refolding of these structures.

It thus appears that the emplacement of the granite overlapped the final stages of cleav-age formation which suggest a Lower Devonian age rather than an Upper Silurian one. These techniques do not give us an 'absolute' date for the intrusion in terms of millions of years, for this can only be obtained by radio-metric methods, but it does give quite a precise age for the granite relative to the other surrounding rocks.

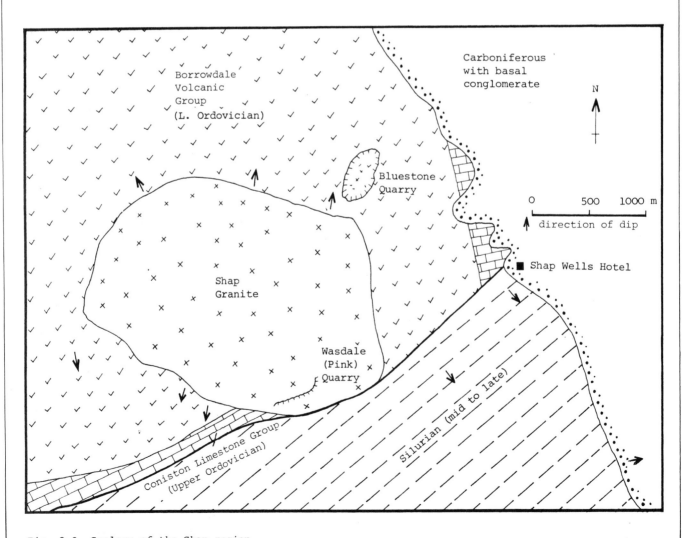

Fig. 2.9 Geology of the Shap region.

The Stratigraphic Column and Stratigraphic Nomenclature

Using the variety of methods open to the stratigrapher, a pattern and order became gradually established during the eighteenth and nineteenth centuries. It was found convenient to divide earth history into a number of sections. For the sake of precision, it became helpful to further subdivide these sections and gradually the familiar stratigraphic column was devised (Table 2.1).

Strictly speaking it is necessary to employ two sets of terms when dividing up the stratigraphic column. One set of terms is based upon the time interval (geological time units) and the other is for the strata deposited within that time interval (time-stratigraphic units) see Table 2.2.

TABLE 2.1 THE STRATIGRAPHIC COLUMN

ERA	PERIOD/SYSTEM	EPOCH	MILLIONS OF YEARS AGO
CAINOZOIC	QUATERNARY	Holocene Pleistocene	
			— 1.8 —
	TERTIARY	Pliocene Miocene Oligocene Eocene Palaeocene	
			— 65 —
MESOZOIC	CRETACEOUS		
			— 140 —
	JURASSIC		
			— 195 —
	TRIASSIC		
			— 230 —
UPPER PALAEOZOIC	PERMIAN		
			— 280 —
	CARBONIFEROUS		
			— 345 —
	DEVONIAN		
			— 395 —
LOWER PALAEOZOIC	SILURIAN		
			— 435 —
	ORDOVICIAN		
			— 500 —
	CAMBRIAN		
			— 570 —
PRECAMBRIAN			
— FORMATION OF THE EARTH —			— 4,600 —

TABLE 2.2 DIVISION OF THE STRATIGRAPHIC RECORD

GEOLOGICAL TIME UNITS

Era	Period	Epoch	Age
Example:		Late	
Mesozoic	Cretaceous	Cretaceous	Campanian

TIME-STRATIGRAPHIC UNITS

(Erathem)	System	Series	Stage
Example:		Upper	
(not often used)	Cretaceous	Cretaceous	Campanian

General usage has meant that some of these terms have come to be applied loosely, yet each has its own quite precise meaning. Perhaps an example of how each set of terms should be applied will illustrate the point. The Jurassic <u>Period</u> is a distinct portion of time which is preceded by the Triassic Period of time and followed by the Cretaceous Period of time. During these periods of time the rocks which comprise the Triassic, Jurassic and Cretaceous <u>Systems</u> were deposited. Thus, it would be correct to say that the Portland Limestone belongs to the Jurassic System but that certain ammonites flourished during the Jurassic Period.

These periods and systems are then further subdivided so that <u>Age</u> is the name given to the portion of time during which the rocks that form a 'stage' were deposited. Stages are usually named by adding the suffix '-ian' to the type-locality where the particular stage was first recognised: eg. Portlandian or Campanian or Ashgillian.

Further subdivisions are possible. First based solely on lithology – <u>rock units</u>. In any exposure of stratified rocks one can see that the rocks are divided into separate layers or beds. The distinctive bedding is due to colour, composition, textural changes and minor breaks in deposition. A number of lithologically similar beds are known as a <u>formation</u> and several formations combine to form a <u>group</u>. The term <u>member</u> is used as another subdivision coming between bed and formation. These rock units are not time-dependent and can only be used within local areas, as environments responsible for particular kinds of sediments often moved laterally with time. That is, deposition may have started in different places at different times and so the junctions are 'diachronous'.

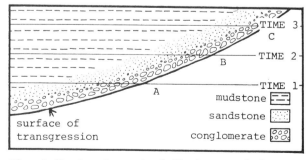

Fig. 2.10 Development of diachronous beds.

Figure 2.10 illustrates how diachronous (literally 'across time') beds can develop. At each time in the case illustrated there are three distinct 'facies'. That is, at the same time in different places different sediments are being deposited. In shallow water a conglomerate develops, in slightly deeper water sand is being deposited, with mud in the deepest water. This pattern of coarse to fine sediments can be seen on a modern sandy beach. Pebbles and coarse debris tend to accumulate at the top of the beach around the high water mark but if you go swimming and wade out beyond the low water mark you will usually be able to feel that the sea bed is muddy.

As the sea level rises, a transgression across the land surface occurs and the different sediments are deposited at different places as time progresses. Thus, at point A the conglomerate is deposited at time 1 but at point B it is laid down at time 2 and so on. Therefore the bed of conglomerate (or for that matter the sandstone or mudstone) is not a continuous bed but it is the result of the same facies being developed at different places at different times.

A biostratigraphic unit is a thickness of strata characterised by one or more diagnostic fossils. The basic unit is the zone. Although such units are not always strictly time-dependent (see the Unit Fossils and Time) they are important in establishing time-stratigraphy and are sometimes regarded as further subdivisions of stages.

The column, as we know it, took many years to 'evolve' and there were a number of disputes about where boundaries between periods should be drawn; in fact some still have not been satisfactorily concluded. One notable dispute arose over the divisions of the Lower Palaeozoic. In the nineteenth century two well-known British geologists were working in different parts of Wales and the Welsh borders. One, Murchison, started at the top of the sequence and worked down whilst the other, Sedgwick, started at the base of the succession and worked his way up. The bottom part of the sequence was called the Cambrian (after the Roman name for Wales) and the top was called the Silurian (after an ancient Welsh tribe). They could not agree where one period stopped and the other started. This rather acrimonious debate was only settled when a third geologist, Lapworth, proposed another period, the Ordovician (named after another ancient Welsh tribe) which covered the disputed time interval. A full table outlining the origin of the names of the periods and the districts where they were first described is given in the Unit Fossils and Time.

As far as possible, the boundaries between the various periods are drawn on significant events that can be recognised over a wide area. This could be a marked unconformity but it is far more likely to be related to the fossil content of the rocks. By careful selection of assemblages of fossils used to mark the boundary, correlation can be made on a worldwide basis.

These divisions of the stratigraphic column will only produce a relative time scale. One period of time can be said to either precede or to follow another but it is not possible to state just how much later in terms of millions of years. However, you will notice that dates are given for the various boundaries shown in the stratigraphic column. These 'absolute' dates have been established using entirely different methods. The dates quoted in Table 2.1 will not be uniformly acceptable to all geologists and, as techniques become better established, they are subject to further refining. How these dates are derived and the problems and sources of error associated with them are discussed in the next section of this Unit.

3
'Absolute Dating' – Non Radioisotopic Methods

Historical Background

As the relative time-scale became established, it was a natural consequence that man would try to assign dates to the beginning and the end of each Period and to the origin of Earth itself. This was not a new problem; for centuries man had wrestled with the question of when he had originated and how old the Earth was. There was a great variety of different estimates for the age of the earth, from a few thousand years to the many millions of years calculated by ancient Hindu astronomers.

However, since the Middle Ages, the story of the Creation as written in Genesis was taken as the literal truth. In fact, very precise times for the origin of the earth were worked out, based upon biblical accounts. In 1644, the Vice-Chancellor of Cambridge University, John Lightfoot, stated that the time of the origin of the earth was 9.00 a.m. on 17 September 3928 B.C. Six years later, the Primate of Ireland, Archbishop Ussher, working along similar lines, arrived at the conclusion that the earth was formed on the night preceding October 23 4004 B.C. This latter date came to have wide acceptance and was even included as a marginal note in bibles published by the Oxford University Press as late as 1910.

This theological constraint of having to compress the history of the earth into 6,000 years had a serious effect on geological thinking. On this time-scale, such features as river valleys could not have been formed by the slow processes of erosion. Instead, the physical features of the earth must have been a result of a series of catastrophic events. Many people may have been rather sceptical of this telescoped time scale, but it was not seriously challenged until the late eighteenth century.

Around his native Edinburgh, James Hutton carefully observed the rocks and realised that many of the features could have been formed by the same processes of erosion and deposition, or volcanic activity, that are evident on the earth's surface at the present time. If these rocks were to have been formed by the slow processes of erosion and deposition, then the earth could not be a mere 6,000 years old, but countless millions of years old. Hutton never tried to put a date on the age of the earth; he just realised that it formed an unimaginably long time ago.

Hutton's ideas, which were later very eloquently written up by John Playfair, Professor of Natural History at Edinburgh University, initiated a school of thought which came to be known as 'Uniformitarianism'. Hutton's followers saw that the natural features of the earth could all be produced, given sufficient time, by similar forces to those acting on the earth's surface today, rather than by a series of catastrophes, as had previously been envisaged. They did not imagine any forces at work in the past that could not be seen in operation on the Earth's surface today. The ideas of Uniformitarianism have often been summed up in the phrase 'the present is the key to the past'.

In 1830, Sir Charles Lyell wrote his very influential book *Principles of Geology* in which he stressed the ideas of Uniformitarianism and it is perhaps more due to Lyell rather than the earlier works of Hutton and Playfair that the uniformitarian ideas gained wide acceptance and finally overthrew the catastrophists. Lyell argued strongly against many of the conventional ideas for the age of the earth and he cited numerous examples to suggest its great antiquity.

There were still those who thought the fossil shells of sea creatures that were found high up in the mountains such as the Alps or Pyrenees were deposited there at the time of the great flood. Lyell pointed out that while a flood may leave behind it, on the surface, a scatter of shingle and mud containing shells of marine creatures, in fact the whole mass of the mountains was composed of the strata containing the fossils.

Other writers in Lyell's day argued that the successive strata that could be seen were once laid down on the bed of an ocean and at the time of the great flood these strata became converted into the land on which we live. Lyell could not accept this version of the creation of the land. He saw that the thousands of feet of deposits, many containing fossils of sea creatures but interbedded with others containing fossils of fresh water or land creatures, represented '... repeated revolutions which the earth has undergone and the signs which the existing continents exhibit, in most regions, of having emerged from the ocean at an era far more remote than four thousand years from the present time.' He also noted that, in the older rocks, the fossils became increasingly divergent from the present day creatures and that these great changes in the nature of animal and plant life inhabiting our planet would not have happened with the speed required to meet the time-scale of the catastrophists.

Fig. 3.1 Fossil Gryphaea, covered both on the outside and inside with fossil serpulae.

To further emphasise the great passages of time needed for the accumulation of the strata Lyell demonstrated that some fossils must have lain on the sea bed for quite a while after death before their eventual burial. The fossil *Gryphaea* (Fig. 3.1) has attached to it a number of serpulae (fossilised worm tubes).

Those on the outside could have developed there during the life of the bivalve but it also has some on the inside, one covering the adductor muscle scar. This could only have grown after the death of the oyster. He also used the example of the fossil echinoid with a single attached bivalve shell. In life, the echinoid had spines which would have prevented the bivalve from becoming attached, so the echinoid must have died and its spines

dropped off before the bivalve became fixed to it. This bivalve died in its turn and one of its shells was separated. All this happened before the echinoid was finally buried in the sediment.

Today, we would not accept that the present is typical of the whole of geological time. For instance, today it is believed that there is far more land exposed than there has been for most of geological time. Nevertheless, in general terms, much of what is happening today must have been happening in the past and the arguments which Hutton and Lyell used to demonstrate the great age of the earth are just as valid today. Uniformitarianism is still a very powerful geological concept.

Lord Kelvin and the Cooling of the Earth

Lyell and Hutton never tried to put an absolute age on the earth, in fact Hutton used the phrase 'no vestige of a beginning – no prospect of an end' as he visualised the cycles of erosion and deposition being repeated over and over again. However, many scientists believed that there must have been a definite beginning and dates were sought for this beginning by a variety of methods.

Although the idea of the earth being only thousands of years old had become unacceptable, there was not unanimous agreement about the length of geological time. Indeed, in 1862, Lord Kelvin made an estimate of the age of the earth which severely limited its duration. Kelvin was a very eminent and influential physicist and he based his estimates on the heat loss from the cooling earth.

It had been noted in deep mines that the temperature increased with depth and this temperature increase was fairly uniform in different parts of the world. Kelvin envisaged the earth originating as a molten planet which had gradually been cooling down. He tried to use the figures obtained from measuring the temperature differences between the top and bottom of the mines to give a cooling rate for the earth. He then estimated the original temperature of the cooling planet and was thus able to produce a figure for the length of

time the earth had been cooling. Because there were many estimates and uncertainties in his equation Kelvin allowed himself a wide margin of error but he concluded that the solid crust of the earth must have been formed between 20 and 400 million years ago. As time went on, he further refined these figures and by 1897 his limits had shrunk to between 20 and 40 million years. This rather short estimate of the age of the earth did not please everyone. Many geologists favoured a much longer time span for, with Kelvin's dates, the geological processes in the past would have to have been a great deal more vigorous. This did not fit in with the Uniformitarian way of thinking.

Kelvin had many detractors amongst geologists. However, such was his prestige that he influenced the work of other people who were also trying, by different methods, to obtain the age of the earth. There was a tendency to obtain figures which fitted in with Kelvin's ideas.

The reason why Kelvin obtained incorrect values is now obvious to us; the earth is in fact not cooling down in the way Kelvin imagined as there is an internal source of heat, namely radioactive decay. Indeed, Kelvin had added the proviso that his figures would only be correct if there was no other source of heat.

Rates of Sedimentation

There were other methods, besides Kelvin's, that were used to try to obtain a figure for the age of the earth, many of these predating Kelvin's attempts.

One method that was used by a number of workers over a period of more than 50 years was to try to determine the age of the earth using the thickness of sedimentary rocks that had been deposited throughout earth history. By obtaining a figure for the rate at which the sediment accumulated it was then possible to calculate the time it would have taken for this sediment to collect.

Table 3.1 illustrates that this method produced a great range of values. This is not surprising when it is realised how many

different estimates have to be brought into the calculation.

One feature which does become clear when looking at the table is that the later workers tended to get lower figures for the age of the earth than the earlier ones, which of course make them more inaccurate. One is left wondering how much their values and calculations were influenced by Kelvin's short time-span for the age of the earth!

The first factor which is open to different interpretations is the actual total thickness of sedimentary strata. A very wide range of values is quoted in the table. The later ones tend to be the greater values, in fact, one even more recent figure has raised the total thickness of strata to 138,000 metres

DATE	AUTHOR	MAX. THICKNESS (in metres) *	RATE OF DEPOSITION (years for 1 metre)	AGE OF EARTH (in millions of years)
1860	Phillips	22,000	4,369	96
1869	Huxley	30,500	3,280	100
1871	Haughton	54,000	28,260	1,526
1878	Haughton	54,000	?	200
1883	Winchell	?	?	3
1889	Croll	3,650	19,680**	72
1890	de Lapparent	46,000	1,968	90
1892	Wallace	54,000	518	28
1892	Geikie	30,500	2,394–22,304	73–680
1893	McGee	80,500	1,968	1,584
1893	Upham	80,500	1,036	100
1893	Walcott	–	–	45–70
1893	Reade	9,650	9,840**	95
1895	Sollas	50,000	328	17
1897	Sederholm	?	?	35–40
1899	Geikie	?	?	100
1900	Sollas	81,000	328	26.5
1908	Joly	81,000	984	80
1909	Sollas	102,500	328	80

* These figures mostly refer to sediments
deposited since the end of the Precambrian.

** Rate of denudation.

(Holmes 1965). The reason for this steady
increase in the values is our expanding
knowledge of successions in different parts
of the world. Many of the early workers in
Europe were using figures for the total
thickness of systems which were nowhere near
the maxima deposited in that system. For
example, the total thickness for the
Cretaceous system in England is less than
1,600 metres, yet in the same time nearly
16,000 metres of sediment were deposited in
California. With many other systems too, the
maximum thickness known to have been deposited
in them has increased.

Even having arrived at the maximum thickness
of sedimentary strata, one is left to question
its value, for no account is taken of
contemporaneous erosion or breaks in
deposition. What of the enormous tracts of
early Precambrian rocks, where evidence for
the thickness of sedimentary rocks is either
missing or has become obscured or obliterated
by later metamorphism? In fact, one of the
greatest obstacles to obtaining a reasonable
value by this method was a failure to
appreciate the vast proportion of earth
history occupied by the Precambrian.

Even more variable than the actual thickness
of the strata is the value used for the rate
of sedimentation (or, in two cases, the rate
of denudation). Accurate rates of sedimentation, over quite long periods of time, have
been obtained in certain localities. It has
been possible to work out the amount of
sediment that has accumulated in the Nile
delta since the reign of Ramases II, over
3,000 years ago. This gives a rate of sedimentation of 1 metre in 1,200-1,500 years.
However, the Nile delta is clearly an example
of an area where sediment accumulates very
rapidly and so this would not be a typical
rate. In fact, most of our known examples of
rates of sedimentation are taken, for
obvious reasons, from inshore environments,
rates for other areas being more difficult
to obtain. Also, one must question just how
far the Doctrine of Uniformitarianism can be
stretched. Would average rates of sedimentation for today be valid for most of
geological time when there was probably far
less land exposed?

For these various reasons, estimates based
on the thickness of sedimentary strata cannot
be taken as typical whilst the wide range of
values obtained illustrates the rather
haphazard nature of this method.

Salt Content of the Oceans

Assuming that the oceans started out as fresh
water, it was argued that it should be
possible to determine their age, and therefore
the age of the earth, by calculating the rate
at which sodium is added to the oceans. In
1899 Joly (an Irish geologist) set out to make
this calculation.

Joly estimated that 160 million tons of sodium
were carried to the sea every year by the
rivers. He then calculated the total volume
of the water in the oceans and multiplied
this figure by the concentration of sodium in
sea water. This gave him a value for the total
sodium content of the oceans. Using the
calculation:

$$\text{age of oceans} = \frac{\text{total amount of Na in sea water}}{\text{amount of Na supplied per year}}$$

he estimated the age of the oceans to be 90
million years. This is obviously far too low
for a number of reasons. First, he assumed
that the rate of run-off and the sodium
content of the rivers had remained constant
throughout geological time. This assumption,
given that the amount of land exposed is
variable, is highly contentious. It is very
likely that the present volume and sodium
content of rivers is well above average due to
the relatively large amounts of land exposed.

Joly also did not make sufficient allowances for
the amount of sodium that does not remain in sea
water because it:
- gets locked into evaporite deposits;
- is blown onto the land in sea spray;
- becomes incorporated into marine sediments.

However, there are so many uncertainties in this method, that even a quite recent calculation, making due allowances for these sources of error, produced a value that could vary by over 1,000 million years, but at the absolute maximum only put the age of the oceans at 2,500 million years old.

Varve Counting

As has been illustrated, it is not possible to use the sequence of sedimentary rocks to obtain a meaningful value for the age of the earth, beyond showing that it is much older than calculated by Kelvin. However, there are local examples where accumulation of sediment can be used to date geological events very accurately.

At the edge of ice-sheets, lakes form, often dammed by moraines, into which glacial melt-waters discharge. Whilst there is rapid run-off during the summer months when the ice is melting, there is very little or nothing entering the lake in winter. During the summer, the coarser material settles to the bottom but the finer material may take until winter to settle from suspension. The effect is to produce a banded sediment or varve, which not only grades in size but also in colour as the finer material is usually darker.

De Geer, a Swede, recognised that this pattern was the product of an annual influx of sediment and that the number of years of sedimentation into a lake could be caculated by counting the varves. De Geer also noticed that there were particularly thin or thick layers which were due to varying annual weather conditions. He was able to use these distinctive layers as marker horizons and, by recognising the same layers in different sections, he was able to correlate over wide areas and so build up a continuous record of annual deposition stretching back into the Pleistocene. Because the top of the section could be linked to a known historical date, accurate 'absolute dates' could be given to Pleistocene deposits. De Geer was able to apply his technique over wide areas. As the ice sheets retreated from Northern Europe, there were periods when they remained stationary for a time, producing terminal moraines. De Geer, counting back varves and correlating from one area to another, was able to be quite precise about the dates when these terminal moraines were formed. Figure 3.2 shows these dates which provide a record of the retreating ice.

Today, there are still two areas where ice-fields linger in the mountains in Scandinavia. De Geer thought that a particularly thick varve in Lake Ragunda marked the time when the two ice sheets split - the so-called Bipartition. By counting back from a known historical point he was able to arrive at a date of 6850 B.C. for the Bipartition and this date was later confirmed by radio-carbon methods.

The varve sequence in Scandinavia provides a unique opportunity to produce 'absolute dates' for geological events by actually physically counting them back. Nowhere else is the sequence complete enough for accurate dates to be obtained. For instance, in North America there are several large gaps in the sequence. There is no reliable estimate of their duration and no valid link has been made between the top of the sequence and a known historical event and so no accurate dating of the events can take place using this technique.

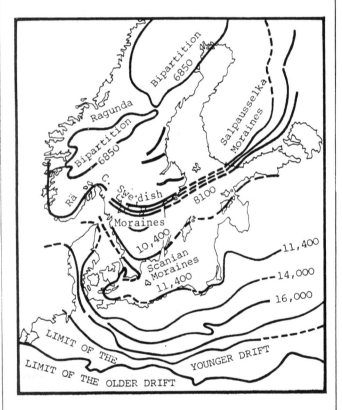

Fig. 3.2 Map of the chief moraines laid down as the ice sheets retreated into the mountains of Scandinavia. The figures date the approximate positions of the successive ice fronts in years B.C.

Growth Bands on Corals

For a long time, growth bands have been noticed on corals, not only modern corals but also well-preserved, fossil ones. At first, these were taken to represent annual growth cycles, rather like the rings on trees, although it was by no means certain, since many other invertebrates were known to have growth cycles which are not a simple annual variation.

A worker in 1963 took a more detailed look at some modern specimens than anyone else had done before him and he came to a surprising conclusion. He interpreted the tiny growth lines as a product of daily growth, as he was able to count about 360 tiny lines in the area the coral was thought to have grown during one year. Corals live in symbiosis (living together for mutual benefit) with

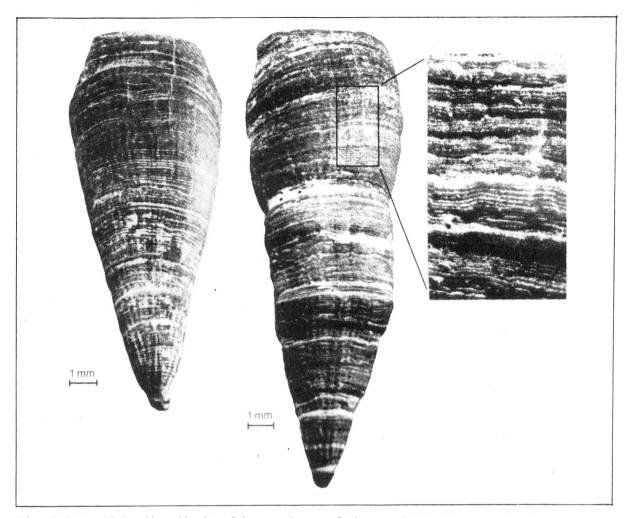

Fig. 3.3 Growth banding displayed by specimens of the coral Holophragma calceoloides *from the Devonian Visby Marl of Gotland.*

algae and as the rate at which the calcium carbonate is produced is thought to be linked to the photosynthetic activity of the algae, which alters between day and night, the idea of daily growth lines is not unreasonable.

On studying specimens of Devonian corals, he found that in every case, the number of smaller lines between larger annual rings was greater than 365, in fact it averaged 400. By a similar examination of some Carboniferous fossils he found that his figures ranged from 385 to 390.

It would have been easy to dismiss this work as having been founded on an incorrect premise, namely that the fine growth lines were not daily increments in an annual cycle. However, it has long been suspected from geophysical evidence that the earth's rotation may be slowing down due to tidal friction. Here was some independent confirmation of this theory.

If an accurate table of number of days in the year was produced it should then be possible to date organisms. For example, a coral showing 393 daily growth bands would be Carboniferous in age (Fig. 3.4). However, as the number of days in a year only changes by one day in ten million years, it is not possible to be completely accurate using this method and it also relies on very good preservation of the specimens and a very patient palaeontologist! The length of the year, of course, remains constant; only the days get shorter

the further back in time we look, e.g. about twenty-one hours in the Palaeozoic.

Despite these problems, here is another possibility where counting the lines (as in varves) may help us to give dates to geological events.

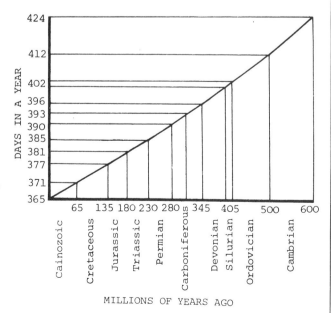

Fig. 3.4 The number of days per year throughout geological time calculated in accordance with the idea that the rate of the earth's rotation on its axis is decreasing as a result of tidal friction.

4
'Absolute Dating' – Radioisotopic Methods

Principles of Radioisotopic Dating

We have already seen that the efforts of the nineteenth century physicists to provide dates for geological events were at variance with the geologists' own evidence (p. 12). Ironically, the very process of heat generation which was unknown to Lord Kelvin was later to provide the basis for the most widespread methods of dating used by modern geologists. The process was that of radioactive decay.

Radioactive decay was first discovered by Henri Becqerel who noticed that photographic plates became fogged when kept in the same cupboard as some specimens of uranium ores. Radioactivity is the process whereby certain elements emit particles and radiation from their nuclei and are thus transformed into other, more stable, elements. Heat energy is produced by these emissions.

From the initial discovery of radioactivity in 1896, rapid progress was made by the physicists in investigating the phenomenon. By 1906 it had proved possible to measure, in the laboratory, the rate at which a radioactive 'parent' element decays to produce a stable 'daughter' element. If, therefore, the ratio of 'parent' to 'daughter' elements could be measured in a rock sample, it should give an indication of its age. This is assuming that there has been no loss of these elements outside the sample being measured and also that allowance can be made for any material of the same composition as the ultimate daughter element, but of non-radiogenic origin.

The earliest measurements were carried out on the more obviously radioactive minerals, such as radium and uranium and in 1911 the first paper giving tentative dates was published by Arthur Holmes. He calculated a maximum date of some 2000 million years for some of his specimens, thereby pushing back Kelvin's original dates by a factor of at least 20. Modern dating techniques use a variety of radioactive elements, which are mostly present in minute quantities in common minerals. A selection of these radiometric techniques, together with their geological applications, will be described later, but first we must consider some basic principles of physics. The following section is intended to give a very simplified version of what happens during radioactive decay and to introduce some of the terms and 'conventional signs' which are used by physicists.

The structure of the atom

In simple terms, the atom may be regarded as a nucleus surrounded by one or more electrons which move around the nucleus in orbits of discrete energy. The nucleus of the atom is composed of smaller particles, namely protons and neutrons. A proton is a positively charged particle, a neutron is a particle of neutral charge and an electron is a negatively charged particle. Atoms are characterised by the numbers of such particles which are present. These are shown by means of two numbers, the mass number and the atomic number.

Mass number: the number of protons and neutrons in the nucleus.
Atomic number: the number of positively charged protons in the nucleus. It is numerically equal to the number of orbital electrons in the neutral atom, since the number of positive and negative particles balance each other out: e.g. for hydrogen (H) it is 1; for lead (Pb) it is 82.

These numbers are shown by conventional placings around the chemical symbol of the element concerned, as, for example, in the form of lead:

$^{207}_{82}Pb$, 207 is the mass number; 82 is the atomic number.

In other words, this form of lead has:
- 82 protons
- 82 electrons
- 207-82 = 125 neutrons

Isotopes: (isos – equal; topos – place) i.e. occupying the same place in the Periodic Table of the elements.

The form of lead quoted above is not the only way in which the element can occur. Other varieties are known, having a different number of neutrons and therefore the same atomic number, but different mass numbers. They are as follows:

	$^{204}_{82}Pb$	$^{206}_{82}Pb$	$^{207}_{82}Pb$	$^{208}_{82}Pb$
Number of protons (& also electrons)	82	82	82	82
Number of neutrons	122	124	125	126

Such varieties of an element are known as isotopes. In the case quoted above only $^{204}_{82}Pb$ originates as 'primeval' lead. All the other isotopes are of radiogenic origin: that is, they are the end products of a radioactive decay series, which started as totally different, unstable elements such as uranium and thorium. Chemically, the isotopes of lead behave in identical ways, since the chemical properties of an element are determined by its number of electrons.

The mechanism of radioactive decay

There are two main mechanisms of radioactive decay which are of relevance to the geologist.

These are as follows:

1. **Loss of an alpha-particle from the nucleus**
An alpha-particle is equivalent to the nucleus of the helium atom and consists of 2 protons and 2 neutrons. The loss of an alpha-particle gives two effects: the loss of four units of mass and therefore a drop in the mass number of 4; a decrease of two in the atomic number. This means that the daughter material now occupies a different place in the periodic table and is therefore a new element.

2. **Loss of a beta-particle from the nucleus**
This may be thought of as a neutron which decays to form an extra proton and an electron. The electron is emitted. Its loss means that there is then a surplus positive charge in the nucleus and so the atomic number increases by 1, again resulting in a new element. The mass number remains the same. ·

Some of the intermediate stages of radioactive decay are very complex, but for geological purposes we omit reference to them and identify each series simply by the original parent element and its stable daughter. The decay series of most value to the geologist are:

$$^{87}_{37}\text{Rb} \rightarrow \ ^{87}_{38}\text{Sr} \qquad \text{(rubidium to strontium)}$$

$$^{40}_{19}\text{K} \rightarrow \ ^{40}_{18}\text{Ar} \qquad \text{(potassium to argon)}$$

$$^{238}_{92}\text{U} \rightarrow \ ^{206}_{82}\text{Pb} \qquad \text{(uranium to lead)}$$

$$^{235}_{92}\text{U} \rightarrow \ ^{207}_{82}\text{Pb} \qquad \text{(uranium to lead)}$$

$$^{14}_{6}\text{C} \rightarrow \ ^{14}_{7}\text{N} \qquad \text{(carbon to nitrogen)}$$

Most of these involve a net loss of α or β particles, although in the case of $^{40}_{19}\text{K} \rightarrow \ ^{40}_{18}\text{Ar}$ the transformation is rather more complex.

The rate of radioactive decay

The breakdown of radioactive elements to other isotopes is a well-ordered process. It happens at rates which can be determined in the laboratory and which appear to be independent of external changes in temperature and pressure, even the greatly increased temperatures and pressures to which rocks are subjected when they are deeply buried. This is why the process has such great potential as a tool for dating rocks.

The decay constant and half-life

The law of radioactive decay is that in any given unit of time, the number of atoms (n) of an element which decay is directly proportional to the number of atoms (N) of the mother element. For each radioactive element, then, the decay constant (λ = lamda) is given by:

$$\lambda = \frac{n}{N} \text{ per year}$$

e.g. for radium, $\lambda = 0.0004273$ per year. Thus 4273 atoms of radium decay every year from an original total of 10,000,000 atoms. Radioactive elements do not decay <u>completely</u> to ·daughter elements; there is always a trace

left, even if it is not possible to measure it. We cannot therefore give a time for <u>all</u> the element to decay, but we can say how long it takes for half of it to go. This time is known as the half-life, and it can be derived from the decay constant by the equation:

$$\text{Half-life} = \frac{0.693}{\lambda}$$

For most practical purposes, geologists usually quote the half-life of an element rather than its decay constant.

<u>HALF LIVES OF SOME GEOLOGICALLY USEFUL
ELEMENTS</u>

^{87}Rb to ^{87}Sr - 48,800 million years
^{40}K to ^{40}Ar - 11,930 million years
^{238}U to ^{206}Pb - 4,467 million years
^{235}U to ^{207}Pb - 704 million years
^{14}C to ^{14}N - 5,730 <u>years</u>

Measurements of the ratios of 'parent' and 'daughter' isotopes are usually made with a mass spectrometer and are beyond the scope of elementary laboratories. It is, however, possible to simulate radioactive decay using readily available 'equipment' such as coins and dice.

Try the following; plot the results and then compare with the 'ideal' half-life curve in Fig. 4.1.

1. *Coin tossing. Toss a large number of coins. Discard any 'tails' and record how many 'heads' are left. Toss again and again until only one or two coins are left in. The 'tails' represent atoms which have decayed. Plot out the results:*
y axis = number of atoms of 'parent element'
x axis = number of throws (i.e. 'time')
Is there a constant 'time' interval each time you halve the number of atoms? Why is this game likely to give a rather less-than-perfect curve?

2. *Dice throwing. Do the same sort of thing with a number of dice, discarding the 'six' each time. Plot out the results. In what ways is the curve similar to/different from the coin curve? What is the 'half-life' of the 'radioactive dice' in terms of 'time' (i.e. number of throws)?*

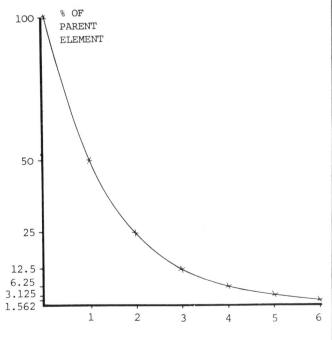

Fig. 4.1 The half-life curve.

In real situations, measurement of the ratio of parent atoms to daughter atoms will give an indication of how many half-lives have elapsed since the parent atoms were formed. If the half-life is known accurately, the age of the material can be calculated.

The Application of Physical Principles to Geological Situations

It is important to realise that, in geological situations, the 'radioactive clock', is not going at the time of <u>crystallisation</u> of a mineral or rock, but rather at the time when it has cooled to its <u>blocking temperature</u>. This is the point at which the mineral or rock becomes a closed system i.e. when transfer of elements between it and its surroundings ceases. For the $^{40}K/^{40}Ar$ decay series the blocking temperature may be as low as 200°C.

Dating igneous rocks

Perhaps the most obvious application of radiometric dating is to igneous rocks, especially those which have cooled quickly. This would include lavas and tuffs, which often occur inter-bedded with sediments. The latter may have been assigned dates on the relative time scale from their fossil content, so here is a possible means of relating dates in years to the well-known geological Periods (Fig. 4.2).

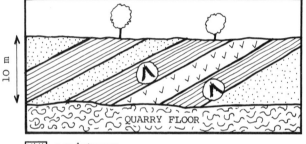

sandstones

shales with graptolites

lava dated at 490 ± 8 m yrs.

Fig. 4.2 'Absolute' dating from a lava flow in an Ordovician sequence.

Minor intrusions such as dykes and sills took rather longer than lavas to cool, but they often provide reliable radio-isotope dates. Again, these dates can frequently be related to fossiliferous strata injected by the minor intrusion and further 'absolute dates' thus assigned to the relative time scale (Fig. 4.3).

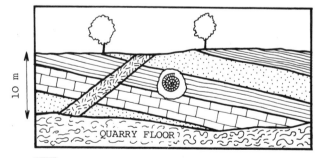

shales with ammonites

sandstone

limestone

dyke rock dated at 140 ± 4 m yrs.

Fig. 4.3 Assigning a limiting date to part of the Jurassic from a dyke. The rocks must be older than 140 million years but in this case no maximum date is obtainable.

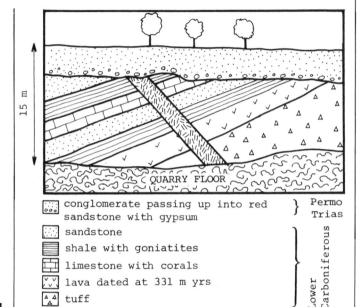

conglomerate passing up into red sandstone with gypsum } Permo Trias

sandstone

shale with goniatites

limestone with corals

lava dated at 331 m yrs

tuff

} Lower Carboniferous

dyke dated at 315 m yrs

Fig. 4.4 A short exercise for you. Assign dates in years as closely as you can to the following: (a) the sedimentary rocks of the Lower Carboniferous; (b) the base of the Permo Trias.

Dating deep-seated intrusions such as batholiths, stocks and bosses is potentially more difficult, since different parts of a large intrusion cool at different rates, depending upon the crustal level reached. Earlier attempts to date large intrusions appeared disappointing, since results were inconsistent, but it is now appreciated that these very inconsistencies are capable of revealing much detail of the cooling history of an intrusion, when they are interpreted intelligently. We shall return to this theme later.

Dating metamorphic rocks

Traces of radioactive minerals are present in most rocks, so if a sediment or an igneous rock undergoes metamorphism the 're-grouping' of its constituent elements may well initiate another 'radiometric clock'. Again, the age is measured from the time at which the new minerals of the metamorphic rock become closed systems to the radioactive isotopes. Many metamorphic rocks contain micas, feldspars and hornblende, all of which may be suitable 'host minerals' for traces of ^{40}K or ^{87}Rb. The application of radiometric techniques to metamorphic rocks is fraught with more difficulties than in the case of igneous rocks, but nonetheless, significant results have been obtained, as we shall see later.

Dating sedimentary rocks

Quite obviously, it is no use extracting mica crystals from a sedimentary rock such as a flagstone, subjecting them to radioisotopic analysis and expecting the results to provide the date of deposition of the flagstone! The micas actually crystallised long before, in some igneous or metamorphic rock, and are present in the flagstone simply because it was produced by deposition of material eroded from the older terrain. Nevertheless, the date of crystallisation of the micas provides a limiting date for the sediment; it cannot be older than its constituent grains! The dates may also be useful in helping to determine the provenance of a sediment and in reconstructing palaeogeographies. For example, micas dated at 280 million years occur in red sandstones and marls in the Midlands, showing that the deposits must be of Permian age or younger. In this case, detailed examination of the mineral content had already suggested that the sediments were derived by erosion from the Hercynian granites of the South-West. This hypothesis for the origin of the red beds is strongly supported by the measured date and also by sedimentary structures which demonstrate a southerly source. Taken together, such evidence has contributed to the palaeogeographic map for the time of deposition of the red beds, which is generally thought to have been in the Triassic Period (about 200 million years ago). See Figure 4.5.

More precise dates for sedimentary rocks may sometimes be derived from radioisotopic measurements of igneous or metamorphic rocks which lie below or above them, or which cut them (Figs. 4.2, 4.3, 4.4). Although most of the minerals with 'dating potential' in sediments are derived from older crystalline rocks, there are some minerals which crystallise in sea water and which, in favourable circumstances, may 'trap' radioactive isotopes. If these minerals remain closed systems to the products of decay, then they can be used to date the time of formation.

The most commonly used mineral of this type is glauconite, a hydrated potassium silicate produced in shallow sea water. Many apparently reliable results have been obtained from glauconitic rocks, particularly of Mesozoic age. Younger rocks are, however, frequently affected by weathering and older ones are altered by depth of burial, so the results are used with caution.

Occasionally, volcanic ash or dust becomes incorporated into sediments. Sediments

containing a very high proportion of volcanic dust are known as 'bentonites' many of which can be dated by radiometric means.

Recently, results have been obtained from clay minerals in some shales, which apparently acquired ^{87}Rb from the sea-water as they grew. This is of importance in seeking the ages of sediments containing evidence of early life and details are given in a later section.

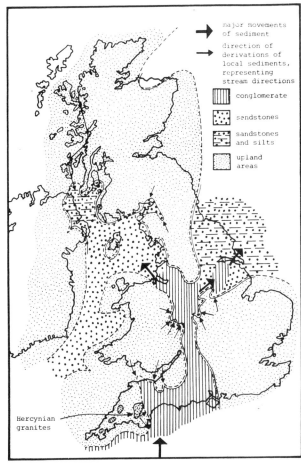

Fig. 4.5 Palaeogeographic map of the early Triassic.

Errors in Radioisotopic Dating

Radioisotopic dates are usually expressed along with a standard error, e.g. 393 ± 11 million years. It is often assumed that the 'plus/minus' figure expresses the geological uncertainties in the measurements, but this is not the case. The standard error figure is simply an indication of the reliability of the laboratory work and expresses the probable limits within which a repeat of the experiment would fall. There may, however, be other problems of a physical nature, or which arise from the geological situation from which the specimens came. These must be borne in mind when the laboratory dates are used to aid interpretation of the geology. The main problems are as follows:

1. Some of the daughter element may have been lost by diffusion or leaching from the host mineral. This would result in an underestimate of the true age.
2. The daughter element may be enriched by processes other than radioactive decay. This is less likely, but if it does occur, the date will be an overestimate.

3. There is some uncertainty about the half-lives of some of the radioactive decay series. This is most noticeable in the $^{87}Rb/^{87}Sr$ series where dates differ by about 6%, depending upon which of two values is used.
4. Dates measured relate to the time when the specimen became a closed system to components of the decay series, not necessarily to the time of crystallisation.
5. Non-radiogenic elements of the same composition as the daughter element may be present in the rock.

In practice, a series of elaborate cross-checks is made and it is often possible to allow for, or rule out, many of the above uncertainties. When the technique was in its infancy, the dates were regarded with some suspicion. Today, however, so many mutually consistent data have been gathered from many parts of the world at a variety of stratigraphic levels that the general validity of the techniques has been almost universally accepted.

Discordant dates

It is, however, widely recognised that there are often discrepancies between results obtained by the use of different techniques applied to the same rock. The same applies to dates derived from minerals separated from a rock compared with those obtained by crushing and testing the rock sample in bulk. At first, such discrepancies led to the virtual discarding of the results, but it is now recognised that these discordant dates, as they are known, can reveal much about the crystalisation history of the rock. For example, a granite may have crystallised from a magma and reached its blocking temperature say 398 million years ago, when the radioisotopic 'clock' would have been set going. If, then, 280 million years ago, the granite suffered metamorphism, the conditions may well have been intense enough for the growth of new minerals and the setting off of another 'radioisotopic clock'. Measurements by several different methods, or measurements on separate minerals compared to the whole rock may well reveal two distinct dates, each of which has an acceptably low experimental error. Such a phenomenon is known as overprinting and it may vary in intensity up to the point where radioisotopic evidence of the original event is obliterated altogether.

Methods of radiometric dating

We shall now examine briefly the main methods of radiometric dating useful to the geologist. These are shown in Table 4.1. The radio-carbon technique ($^{14}C - ^{14}N$) is considered separately since it is of value only in the very limited, topmost parts of the geological column and some of the assumptions underlying its use are rather different.

TABLE 4.1 AN OUTLINE OF THE MAIN METHODS OF RADIOMETRIC DATING.

METHOD (WITH HALF-LIFE IN BRACKETS)	$^{87}Rb - ^{87}Sr$ (48,800 m yrs)	$^{40}K - ^{40}Ar$ (11,930 m yrs)	$^{238}U \quad ^{206}Pb$ (4,467 m yrs)	$^{207}Pb / ^{206}Pb$ RATIO
AGE OF MATERIAL WHICH CAN BE DATED	Older than 10 million years,	Older than 100,000 yrs. Some younger material dated with less confidence.	Older than 20 million years	Over 400 million years.
SOURCE ROCKS AND MINERALS	Muscovite, biotite, all potash-feldspars, glauconite. Whole-rock analysis of granitic gneisses and other metamorphic rocks.	Muscovite, biotite, hornblende, pyroxenes, plagioclase feldspars, high temperature potash-feldspars (but NOT orthoclase), glauconite. Whole-rock analysis of most volcanic rocks, some granites and fine-grained micaceous metamorphic rocks.	Uraninite and pitch-blende in uranium ores. Zircon and monazite in granites.	as left
GEOLOGICAL APPLICATIONS	Dates of original cooling and uplift of acid igneous rocks. Dates of meta-morphic episodes. Dates of deposition of glauconite-bearing sediments. Possible that ratio of other strontium isotopes may indicate ultimate source of granitic magma	Similar to those left.	Dates of crystallisation of uranium ores and cooling of granites. Detecting discordant dates as left.	Particularly useful for dating the earth's oldest rocks and for dating meteorites.
NOTES	1) Uncertainty over the precise half-life of rubidium could result in up to 6% error. 2) ^{87}Sr may diffuse at temperatures of 200-500 C from certain minerals into the bulk of the rock. 'Whole-rock' date therefore represents date of cooling. Separate mineral dates may be those of later metamorphism. 3) Non-radiogenic ^{87}Sr in the original rock must be allowed for. Obtained from ratio of ^{87}Sr to non-radiogenic ^{86}Sr.	1) The daughter-element argon, being a gas, may diffuse out of the crystal lattice and result in an underestimate of the age. Most noticeable in coarse-grained minerals which took a long time to cool to the blocking temperature (the point at which the mineral becomes a closed system to argon loss). 2) More rarely, extraneous argon may diffuse into the system, resulting in an overestimate of the age.	1) Diffusion of some decay products in the series, e.g. radon, can result in underestimates of age. 2) Uranium ores are un-common. 3) The method is less used now than formerly.	This method does not itself use a radio-isotopic decay series. Instead, the ratio of ^{207}Pb produced by the decay of ^{235}U is plotted against ^{206}Pb from decay of ^{238}U, which is about six times as slow. The method cannot be used on geologically 'young' material.

The Radio-Carbon Method ($^{14}C - ^{14}N$)

When the radio-carbon technique of radio-metric dating was first developed, in the early 1950s, it rapidly caught the attention of the general public. The technique was applied to many archaeological discoveries and enabled dates in years to be assigned to the 'stratigraphy' which had already been established for the layers containing evidence of Man's past activities. The technique particularly fired the public imagination when it was applied to the infamous Piltdown skull.

This clever forgery had been 'found' in 1912 and had at first been accepted as genuine. In 1949, however, the remains were tested for their fluorine content, as were the Pleistocene fossils with which they were supposed to have been associated. There was a big discrepancy, indicating that the skull had absorbed far less fluorine from groundwater than the fossils and must therefore be very much younger.

When radio-carbon techniques became available, measurements were soon carried out on the specimens. The date of the skull itself was found to be about 1330 A.D. and the associated jawbone (which actually came from an orang-utan) was dated at about 1450 A.D.

However, the technique has also been heavily criticised on the grounds of reliability and so it has again been brought before the public eye. The result is that the average layman has heard of the technique, but remains blissfully ignorant of the other methods which have just been described. It is therefore commonly assumed that <u>all</u> geological dates are derived from radio-carbon measurements, but nothing could be further from the truth. The half-life of the radioactive isotope of carbon (^{14}C) is only 5730 <u>years</u>, and after about 9 half-lives have elapsed the quantities of ^{14}C remaining in the specimen are so low that measurement is virtually impossible. Thus the maximum age which can be reliably measured by the method is about 50,000 years. This is of considerable value to the archaeologist, and to those geologists interested in the younger part of the Pleistocene epoch, but is not applicable to the rest of the geological record. Even anthropologists working on the possible ancestors of Man have to use other methods to date their material, much of which comes from the older Pleistocene rocks and from the Tertiary System.

The principles of the method

Unlike the other methods, radioactive carbon is not involved with the processes of crystallisation but, instead, is produced in the atmosphere, continuously, by the action of cosmic rays. In effect, the cosmic rays bombard atoms of nitrogen (^{14}N) present in the atmosphere and occasionally dislodge a neutron from the nucleus. Some of these then collide with other nitrogen atoms to produce radioactive carbon, ^{14}C. The number of such transformations is very small, so that ^{14}C only forms about 0.0000001% of the total carbon in the atmosphere. As soon as it is produced, the ^{14}C begins to decay back to ^{14}N by loss of a β-particle at the rate stated

above. It is, however, constantly being replenished by cosmic ray activity and evidence suggests that the balance between ^{14}C and non-radiogenic carbon has been maintained for many centuries, although Man's more recent influence on the composition of the atmosphere has now also to be taken into consideration.

The significance, for dating, is this. Plants and animals absorb carbon dioxide from the atmosphere during their lifetimes. The proportion of ^{14}C to non-radiogenic carbon in their tissues will be the same as that of the atmosphere. However, once the organism dies, the intake of carbon ceases and the ^{14}C begins to decay to ^{14}N without replenishment (Fig. 4.6). The ratio of ^{14}C to non-radiogenic carbon will, therefore, become less and less in comparison to that in the atmosphere and the measurement can be used to find the date of death of the organism in years before the present day (shown as 'years B.P.').

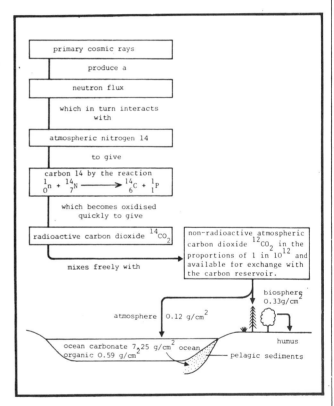

Fig. 4.6 Sketch of the radiocarbon production in the atmosphere and its fixation in the biosphere and other carbon reservoirs. The amount of carbon in different parts of the exchange reservoir is shown, expressed as grams of carbon in exchange equilibrium with the atmospheric carbon dioxide for each square centimetre of the earth's surface. The main reservoir is the ocean carbonate.

Quite a range of materials may be chosen for radio-carbon dating, including wood and ash from archaeological sites, scraps of fabric, shell and bone. Not all the material need be preserved on land; recently the method has been successfully applied to dating younger Pleistocene and Recent sediments from the sea-floor, providing dates for some late-glacial and post-glacial events.

Errors in the radio-carbon method

Before blindly accepting any particular radio-carbon date at face value we must examine some of the assumptions which have been made and look for possible sources of error. One assumption, which is very difficult to check, is that the cosmic ray flux has remained much the same for many thousands of years. It is this cosmic ray activity which is responsible for constantly regenerating the radio-carbon, so if it has varied in intensity the balance of ^{14}C to ^{12}C would have varied too.

There may also have been variation in cosmic-ray flux from one part of the world to another, so making it difficult to compare radio-carbon dates from different regions.

Another problem is that during the last century or so, Man himself has altered the balance of carbon isotopes in the atmosphere. The most noticeable effect has been to increase the output of non - radiogenic carbon by burning 'fossil' fuels, although this is offset, to some extent, by man-made nuclear radiation. Recent work, therefore, has used the carbon ratio computed from wood of 1850 as a reference point, in preference to the ratio of carbon in the modern atmosphere.

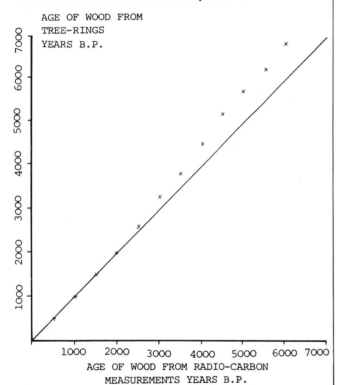

AGE OF WOOD FROM TREE-RINGS YEARS B.P.

AGE OF WOOD FROM RADIO-CARBON MEASUREMENTS YEARS B.P.

Fig. 4.7 Tree-ring calibration of radio-carbon ages.

To some extent, a check may be made by comparing dates derived from counting annual growth-rings on very old trees with radio-carbon dates made at intervals through the same tree. This has been done from specimens of the Bristlecone Pine in the western USA. A simplified version of the results is shown in Fig. 4.7 which shows that there is close agreement for the last 2,500 years or so, but that discrepancies occur in older material. The distance of a tree-ring date above the straight-line graph is a measure of the error

and it shows that the radio-carbon date may be an underestimate by as much as 800 years in 6000 years. Such information is now being widely used to correct radiocarbon dates of less than 6000 years or so.

A further problem is that specimens are particularly susceptible to contamination by ground-water carrying dissolved modern carbon. The result would be an increase in the $^{14}C/$ ^{12}C ratio and hence an underestimate of the true age. The reverse difficulty is encountered when contamination is by 'old' carbon. For these reasons, great care is necessary when collecting specimens and microscope analysis is usually necessary to determine their post-depositional history. Even then, it is not always possible to eliminate the occasional 'rogue' result.

After such a list of potential pitfalls, it may seem surprising that the method has any validity! In fact, however, many thousands of mutually consistent results have been obtained. Some of these can be checked against historical evidence, as part of the following table shows, and the technique is regarded as a valuable tool in dating arch-aeological discoveries and in late Pleistocene geology.

TABLE 4.2 A VARIETY OF APPLICATIONS OF RADIO-CARBON DATING, WITH HISTORICAL DATES WHERE KNOWN.

SAMPLE	EXPECTED AGE YEARS B.P. (1960)	RADIO-CARBON AGE IN YEARS B.P.
Oak from Viking ship (A.D. 800).	1,160 + age of timber	1,190 ± 60
Wood from coffin (A.D. 698) of St Cuthbert, Durham Cathedral	1,270 + age of timber	1,333 ± 150
Carbonised bread from Pompeii, A.D. 79 eruption of Vesuvius.	1,800	1,830 ± 50
Wood from Roman ship of the Emperor Caligula, A.D. 37-41.	1,920 + age of timber	1,980 ± 70
Last hundred rings of a giant Redwood, California.	2,878-2,978	3,005 ± 165
Antler from red deer, Stonehenge, Wiltshire.	3,510-3,560	3,670 ± 150
Vegetation carbonised by the most recent lava flow from Puy de la Vache, Auvergne.		7,650 ± 350
Vegetation carbonised by the eruption responsible for the caldera of Crater Lake, Arizona.		8,060 ± 250
Charcoal in pumiceous ash erupted during the formation of the Laacher See.		10,800 ± 300

In the geological context, perhaps the most useful application has been in establishing a time-scale for events of the last glaciation. Measurements have been made on peat layers and other land-based material, and also on shells in cores from the sea-bed. The results of many different measurements show a rather consistent date for the end of the last Ice Age of about 11,000 years B.P. The previous estimate, based on non-radiometric methods, was about 20,000 B.P., so our ideas about the length of post-glacial time have had to be considerably foreshortened!

5
Case Studies

The Granitic Intrusions of the Lake District

The Shap Granite

In an earlier section (p. 7) we referred to the dating of the Shap granite by the application of geological principles. It was shown that the intrusion is of Devonian age, probably being produced fairly early in the Period. The granite has also been dated by radiometric means, which has yielded a date of 394 ± 3 million years. The value of such a date is twofold: it can be treated in isolation, simply to know the date of the intrusion and it can be collated with many other dates, derived from similar geological situations all over the world, with the purpose of assigning dates in years to the stratigraphic record. Allowing for the fact that the radiometric date represents the time when the granite reached its 'blocking temperature', not that of the original injection, these studies of the Shap intrusion would suggest a date of rather more than 394 million years for the base of the Devonian. Many similar measurements have been made elsewhere in the world and the base of the Devonian has generally been agreed to be about 395 to 400 million years old.

When an igneous intrusion is used in this way for dating purposes, it is referred to as a 'bracketed' intrusion: that is, the radiometric date provides a minimum age for the country rocks into which it is intruded and a maximum age for the base of the overlying succession.

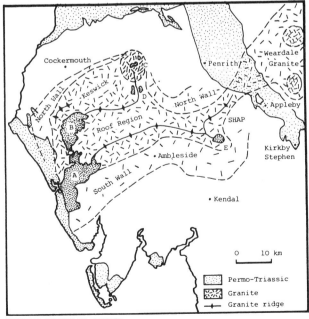

GRANITES EXPOSED AT SURFACE
A Eskdale Granite   B Ennerdale Granophyre C Skiddaw Granite
D Threlkeld Microgranite   E Shap Granite

Fig. 5.1 Sketch map showing the roof and wall regions of the postulated Lake District Batholith and its connection to the Weardale Granite.

Other granitic rocks of the Lake District

The Shap granite is well known because of its distinctive appearance and widespread distribution of glacial erratics, but it really forms a rather insignificant outcrop, when compared with the other granitic intrusions of the Lake District (Fig. 5.1). These comprise the Eskdale granite, the Ennerdale granophyre and the several outcrops of the Skiddaw granite and Threlkeld microgranite.

The question arises as to whether all these outcrops resulted from completely separate magmatic events or whether they represent offshoot stocks from the same major batholith. Several lines of enquiry are possible.

Petrological affinities

All the intrusions are 'acidic', but there is considerable variation between them. The Ennerdale intrusion is a granophyre, i.e. it is a medium-grained rock with quartz/feldspar intergrowth texture. The Threlkeld microgranite is also a medium-grained rock, probably intruded as a laccolith. The others are varieties of granite (using the term in the broad sense), with the Shap granite being markedly porphyritic.

Chemical affinities

Chemical analyses have been made to determine the ratios of sodium, potassium and calcium in each rock and, similarly, the ratios between iron, magnesium and the alkali elements (Na and K together). On both counts, the granites from Shap, Skiddaw and Eskdale plot close together. There are also close petrological and chemical similarities between these granites and the borehole samples of the buried Weardale granite. However, the Ennerdale granophyre is markedly more sodium-rich.

Gravity survey

Granite is of low density compared with that of most country rocks, so gravity surveys over granites commonly show negative readings (or anomalies - see the Unit Geophysics). The negative values continue above buried granites and it was a gravity survey which ultimately led to the discovery of the Weardale granite beneath the North Pennines. An extensive gravity survey of the Lake District has revealed several negative anomalies and the interpretation of the sub-surface geology in Fig. 5.1 is based on these gravity data. It shows the existing outcrops as 'stocks' rising from a large buried mass, connected by a narrow ridge of granite to the Weardale granite. The total mass of granite is certainly large enough to qualify for the name of batholith, although the gravity results reveal nothing about the date of cooling.

Radioisotopic dating

The various radioisotopic studies which have been carried out on Lake District intrusions provide a most interesting example of how careful we must be to interpret the resulting dates in their correct geological perspective. The first apparently reliable dates were obtained by the K/Ar method in the 1960s and early 1970s. These are as follows:

Shap granite	393 ± 11 m yrs
Skiddaw granite	392 ± 6 m yrs
Eskdale granite	383 ± 2 m yrs (measure- ments from an 'atypical' granite exposure)
Ennerdale granophyre	370 ± 20 m yrs

The standard errors of the measurements (except for the Ennerdale intrusion) are quite low, which indicates that we can be reasonably sure of the accuracy of the laboratory work. It is tempting to conclude that the cooling of the various igneous bodies took place over an interval of some 20 million years, with the intrusions of the eastern Lake District being slightly older than those of the west. It was, however, also noted that some of the intrusions had been affected by a later metasomatism (a form of metamorphism) and that this might have altered the argon content and hence the K/Ar date. Many geologists, therefore, regarded the results with some suspicion.

In the late 1970s it became possible to carry out a further series of radioisotopic measurements using the Rb/Sr method. The results are listed below:

Shap granite	394 ± 3 m yrs
Skiddaw granite	399 ± 8 m yrs
Eskdale granite	429 ± 9 m yrs
Ennerdale granophyre	421 ± 8 m yrs
Threlkeld microgranite	439 ± 9 m yrs

In the Rb/Sr method, the ratio between parent and daughter atoms is less susceptible to alteration by later events such as the metasomatism mentioned above. It is therefore generally felt that this second suite of measurements is a more reliable indication of the dates of cooling of the granitic plutons than the earlier set. If this is so, it now seems that the cooling was spread over a longer time interval than was previously supposed, namely some 45 million years. The earlier determination of the age of the Shap granite is confirmed by the more recent measurement, but the other intrusions are now seen to be older, some of them dating back to the mid-Ordovician.

Summary

The several lines of evidence quoted above indicate to most geologists that the surface outcrops of the granitic rocks of the Lake District form part of a batholith which still remains largely buried. The gravity anomalies are also indicative of an underground connection to the Weardale granite (dated from borehole samples at 410 ± 10 million years). At first sight, the range of 45 million years between the oldest and youngest intrusions might suggest that they resulted from quite separate igneous events. However, sufficient work has been carried out on similar igneous complexes elsewhere to show that it often took several tens of millions of years for a batholith to cool. Thus, the Lake District batholith is no different from many others in the world.

Presumably, the broad tectonic setting of the area remained much the same during this time span, as successive offshoots of magma rose into the higher levels of the crust. It has been pointed out that the initial $^{87}Sr/^{86}Sr$ ratios of many of the Lake District samples are in agreement with the kind of values one would expect at a subduction zone. Add to that the style of the deformation of the Lower Palaeozoic rocks and the andesitic nature of the Ordovician volcanics of the Lake District and we have a reasonably clear picture of a destructive plate margin of the Caledonian orogeny.

A 'side-effect' of this recent work has been the suggestion that our previous ideas on the length of Silurian and Ordovician time may be in error. The field relationships of the various intrusions with regard to the stratigraphy and structure of their Lower Palaeozoic 'envelope' are known quite precisely. Hence, the production of new radioisotopic dates for the intrusions means that the dates of the boundaries of the Ordovician and Silurian and possibly the base of the Devonian too, may have to be revised. The most recent papers suggest that the Ordovician may have lasted for 70 million years ('accepted' figure 65 million years) and the Silurian for only 10-15 million years ('accepted' figure 40 million years). It would be ironic if this were the case, since in the early days of stratigraphy, Murchison and Sedgwick established the Cambrian and Silurian Systems, but the Ordovician System was not recognised at all! It remains to be seen whether or not some of these ideas, derived from the geology of the Lake District, achieve more widespread acceptance when compared with dates from the rest of the world.

The History of the Lewisian Complex

We have already noted (p. 18) that it is possible for radioactive material to be redistributed in a sequence of rocks if substantial reheating has occured. Any original radioisotopic date is often 'overprinted' by the later event and may be either completely obliterated or may result in a 'mixed' date being obtained. Such reheating takes place during episodes of metamorphism and it is difficult enough to interpret the dates after just one metamorphic event. Imagine the task where more than one phase of metamorphism is involved!

Just such a situation existed in the Precambrian Lewisian Complex in the North-West Highlands and Islands of Scotland. The Lewisian Complex consists mostly of gneisses (Fig. 5.2) and it has been known for several decades that they have suffered more than one phase of metamorphism. The initial geological mapping and laboratory work was followed several years afterwards by radioisotopic measurements, which confirmed the conclusions reached from purely geological reasoning. This might sound impossible in a region exhibiting such high grades of meta-

Fig. 5.2 *Banded gneisses of variable composition. Scourie, North-West Scotland.*

morphism but, luckily, there are several areas where the later metamorphic events did not wipe out the evidence of the earlier ones, thus providing 'windows', so to speak, into the more distant past.

The map (Fig. 5.3a) shows some of the details of the coast of the North-West Highlands. The older suite of rocks is known as the Scourian, after the village of Scourie and the younger, the Laxfordian, after Loch Laxford. Apart from a tract of country some 55 km long between Lochs Inver and Laxford, the Scourian only 'peeps through' the later Laxfordian in a number of small, isolated outcrops. The

relationships and contrasts between the two suites of rocks are well displayed in the Loch Laxford district itself (Fig. 5.3b). The approximate limit between them is marked by the line XY.

A characteristic feature of the Scourian gneisses, (south of XY) is the presence of anhydrous minerals such as pyroxenes. The dominant trend of the foliation of these gneisses in the Scourie district is NE – SW (Fig. 5.3b).

In contrast, many of the Laxfordian gneisses (north of XY) contain hydrous minerals, such as biotite and hornblende and the main structural trend in this area is WNW-ESE. The intensity of deformation is also greater (see section A-B). The presence of the hydrous minerals suggests that the Laxfordian gneisses were produced by retrograde metamorphism of the Scourian. This may indicate that the temperatures reached were not as high as during the Scourian phase, but the conditions would certainly have been extreme enough to result in overprinting of any radioisotopic dates.

Perhaps the most noticeable evidence that there was more than one metamorphic phase in the area comes from a group of NW – SE trending basic dykes. South of the line XY, these cut cleanly across the gneisses, but to the north of the line they become progressively more deformed and metamorphosed, until they are almost indistinguishable from the host gneisses (Fig. 5.3b).

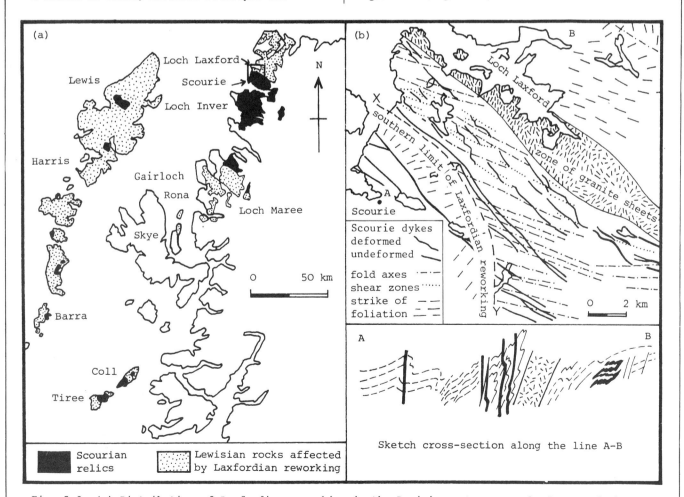

Fig. 5.3 *(a) Distribution of Laxfordian reworking in the Lewisian outcrops on land west of the Moine Thrust. (b) Map and section of the Scourian/Laxfordian boundary (line xy) in the Loch Laxford area.*

SCOURIAN	DOLERITE DYKE SWARM	LAXFORDIAN
Metamorphism and migmatisation of deep levels in the crust	Intrudes Scourian along parallel fractures formed at high level in crust	Earth movements, metamorphism and migmatisation transform Scourian and dykes at moderate levels in crust

  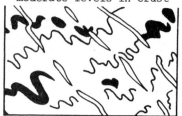

| 2,700 million years | 2,200 million years | 1,800 million years |

Fig. 5.4 *Three stages in the history of the Lewisian.*

The two geologists who were responsible for unravelling the story were Professors H.H. Read and Janet Watson. Needless to say, the situation is far more complex than outlined above, but their simplified diagram of the sequence of events is a useful summary (Fig. 5.4).

In due course, the techniques of radioisotopic dating were applied to the Scourian and Laxfordian gneisses and to the intervening basic dykes. The dating confirmed the general sequence of events which had already been worked out and also showed that each phase of metamorphism covered a large time span, in some cases lasting for 200 million years or so.

The oldest date so far recorded from the Highlands is about 2900 million years. This precedes the main Scourian phase which is thought to have occurred between 2800 and 2700 million years ago. Some geologists believe that this was followed locally by further metamorphic events continuing up to almost 2200 million years ago.

Where the Scourian dykes are undeformed by later events, they reveal a date of 2200 million years following very closely upon the waning phases of the Scourian metamorphism.

In the Laxfordian terrain, two different groups of dates are obtained. Whole-rock determinations by the Rb/Sr method and U/Pb measurements on zircon crystals within the gneisses give a date of about 2700 million years. This is no different from the Scourian results. However, whole-rock dates by the K/Ar method and measurements on separated minerals other than zircon reveal dates of 1900 - 1700

million years ago. This must mean that partial overprinting occurred in Laxfordian times because of reheating of the original Scourian material. Argon diffusion took place, both from individual minerals and from the whole-rock and so the K/Ar decay series was reset. The Rb/Sr decay series, however, remained unaffected and reflects the earlier date.

This shows, quite clearly, that the Laxfordian gneisses were formed by 'reworking' of the earlier material of the local continental crust and were not produced by fresh additions to it from differentiation of the mantle. A summary of the main episodes is given in Table 5.1.

TABLE 5.1 DATES OF LEWISIAN EVENTS.

MILLIONS OF YEARS	MAJOR EVENTS	
1300		LAXFORDIAN
1500	uplift and cooling	
1700		
1900	Laxfordian metamorphism	
2100		
2300	intrusion of Scourian dykes	
2500		SCOURIAN
2700	main Scourian metamorphism	
2900	oldest gneisses in Scotland	

How Old Is Life?

One of the most fascinating problems facing geologists has been the question of the origin of life. Ever since the significance of fossils was first appreciated, we have wanted to know how far down the geological column the record of ancient life extends and to try to derive some idea of the actual dates at which those organisms existed. At one time, fossils were only known from the base of the Cambrian up to the Recent past, a period of time known as the Phanerozoic meaning 'evident life'. Any rocks older than the Cambrian became known as the Precambrian, a logical but rather crude way of labelling nearly seven-eighths of the Earth's history!

In the late 1950s undoubted fossils were discovered in rocks of late Precambrian age and since that time many other occurrences have been noted, some of them in rocks of earlier Precambrian date. Many of these ancient fossils are of microscopic organisms such as bacteria and algae. It should, however, be noted that some workers dispute that the oldest finds are fossils at all, regarding them as being of inorganic origin, or as having been introduced into the microscope slides by modern contamination. Nonetheless, the increased rate of discovery of such fossil claims has promoted much refinement in the use of radioisotopic and other dating methods.

The task is, if anything, more difficult than in the case of igneous and metamorphic rocks, since fossils normally occur in sedimentary rocks which do not lend themselves so readily to radiometric techniques. It has to be said that the geological record has given us no more certain knowledge of the ultimate origin of living cells than it had for James Hutton in the late eighteenth century, or Charles Darwin in the nineteenth.

It is, however, interesting to review the progress which geologists have made so far and we shall examine two examples taken from southern Africa and Australia which together with Canada and the USSR, have provided the best evidence to date of Precambrian life forms.

The Swaziland sequence: South Africa

The Swaziland Sequence consists largely of volcanic rocks but at certain horizons there are sedimentary strata in which the discoveries of micro-organisms have been made. The whole sequence has been involved in some metamorphism, but it was not sufficiently intense to destroy the fossil evidence, nor apparently to overprint a new radioisotopic date. A stratigraphic column is shown in Fig. 5.5 and a location map in Fig. 5.6.

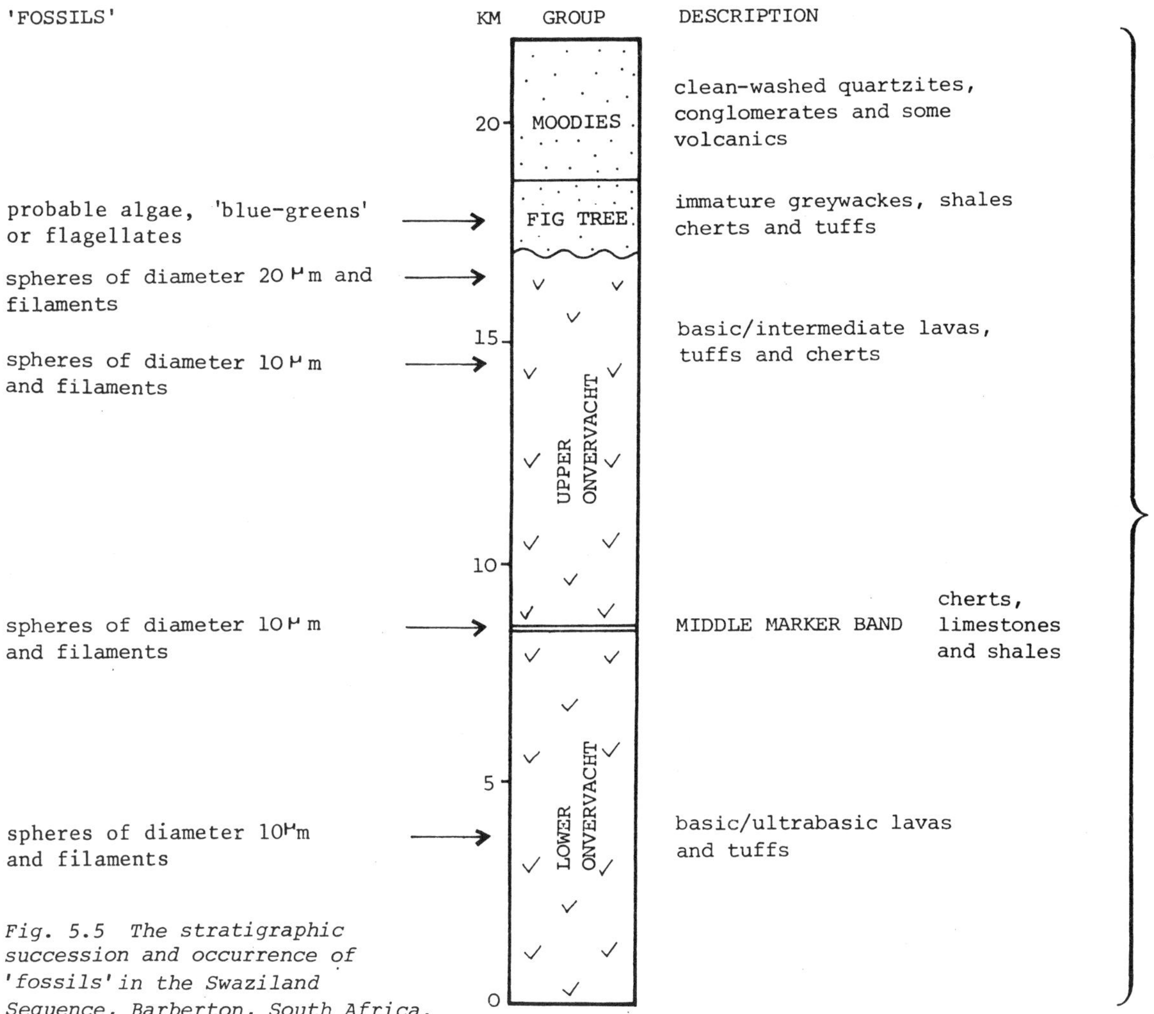

'FOSSILS'	KM	GROUP	DESCRIPTION
	20 —	MOODIES	clean-washed quartzites, conglomerates and some volcanics
probable algae, 'blue-greens' or flagellates →		FIG TREE	immature greywackes, shales cherts and tuffs
spheres of diameter 20 μm and filaments →	15 —	UPPER ONVERVACHT	basic/intermediate lavas, tuffs and cherts
spheres of diameter 10 μm and filaments →			
	10 —		
spheres of diameter 10 μm and filaments →		MIDDLE MARKER BAND	cherts, limestones and shales
	5 —	LOWER ONVERVACHT	
spheres of diameter 10 μm and filaments →			basic/ultrabasic lavas and tuffs
	0 —		

SWAZILAND SEQUENCE

Fig. 5.5 The stratigraphic succession and occurrence of 'fossils' in the Swaziland Sequence, Barberton, South Africa.

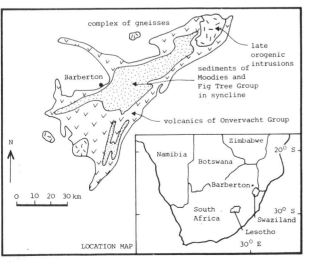

complex of gneisses

late orogenic intrusions

sediments of Moodies and Fig Tree Group in syncline

volcanics of Onvervacht Group

Barberton

N

0 10 20 30 km

LOCATION MAP

Zimbabwe
Namibia
Botswana
Barberton
20° S
South Africa
Swaziland
Lesotho
30° S
30° E

Fig. 5.6 The geology of the Barberton area in Southern Africa.

The lowest, most consistent horizon where radiometric dating can be attempted is the Middle Marker Band. Here $^{87}Rb/^{87}Sr$ techniques were applied to well-laminated shale specimens and a very consistent result of 3355 ± 70 million years was obtained. It seems unusual to try to date a shale by this method, but it is claimed that the rock consists mainly of authigenic clays and that the content of detrital minerals is very low. An authigenic clay is one where the clay minerals formed in the sea water, absorbing as they did so both potassium and radioactive rubidium 87, thus setting going the 'radioactive clock'. The research workers were able to use the ratios of radiogenic and non-radiogenic strontium in modern sea-water as a 'control' and felt confident that the figures obtained represented the date of original sedimentation rather than some later event.

Confirmation that the date is at least approximately correct has come from a variety of sources:

1. The Swaziland Sequence is cut by late orogenic granites and pegmatites of 3070 ± 60 million years (whole rock $^{87}Rb/^{87}Sr$). The sequence must, therefore, be older than 3,070 million years.
2. $^{87}Rb/^{87}Sr$ measurements on shales in the Fig Tree Series gave a good concordant date of $2,980 \pm 20$ million years but the initial $^{87}Sr/^{86}Sr$ ratio suggests that some migration of ^{87}Sr may have occurred, in which case the date may be a slight under-estimate.
3. Measurements of radiogenic lead ratios in quartz-porphyries of the Upper Onvervacht Group resulted in a date of 3400 million years (no error quoted).
4. Uranium/lead ratios in zircons extracted from Onvervacht lavas showed 3,360 million years (no error quoted).

The results of several radiometric techniques, therefore, combine to give dates which are reasonably consistent for these ancient rocks. Unfortunately, geologists are divided about the origin of the 'micro-organisms' in the Middle Marker Band. Some claim that they were indigenous to that horizon; others believe that they were introduced later by ground water, whilst one school of thought argues that they were not produced by living organisms at all. In this case, the reasoning is based on differences in 'organic' chemistry and ratios of $^{13}C/^{12}C$ (not ^{14}C, the radio-active form of carbon).

There is less disagreement about the biological origins of microstructures in the overlying Fig Tree Group, where chemical evidence suggests that photosynthesis had been taking place. The fossils here resemble plant material such as algae, 'blue-greens' and flagellates.

Further north, in Zimbabwe, undoubted stromatolites have been found in rocks of 2,900 to 3,200 million years old, dated by similar means, so quite clearly, there was life in the early Precambrian, even if we are still hazy as to its ultimate origin.

The Ediacara fauna: South Australia

Several discoveries of fossils have been made in the younger Precambrian rocks and it is clear that some of these provide evidence of former animal life. Unlike most of the fossils from the Cambrian Period onwards, these Precambrian creatures have no readily preserved hard parts and so they are only known from imprints preserved under exceptionally favourable conditions. The best known locality for these first Precambrian animals is at Ediacara, in the Flinders Range of South Australia. Similar fossils are known from several widespread localities, including other parts of Australia, Charnwood Forest in Leicestershire, South West Africa, Scandinavia and the USSR, (Fig. 5.7). Representatives of the fauna are shown in Fig. 5.8. They include: possible annelid worms (b); jellyfish (d); colonial coelenterates (e); several organisms of unknown affinities (a), (c), (f), (g).

Fig. 5.8 A selection of fossils from the Ediacara fauna.

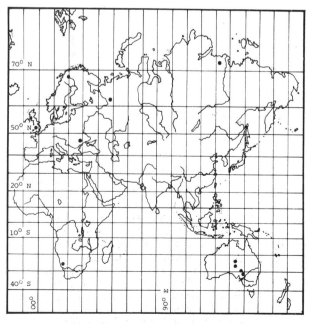

Fig. 5.7 Distribution of localities at which elements of the late Precambrian Ediacara fauna have been found.

TABLE 5.2

LOCATION	AGE (MILLION YEARS)	DATING METHOD AND NOTES
Ediacara (and 3 other Australian sites)	Uncertain, but definitely > 570	Fossils occur in former clay lenses within shallow-water marine sandstones. Radiometric dating is not possible. Some 500 m of mostly unfossiliferous sandstone come between the fossiliferous horizon and the base of the lower Cambrian (570 million years) which is itself unconformable.
Charnwood Forest, England	About 680 $\pm$ 29	Ashy sediments containing *Charnia* are cut by syenites and other igneous rocks about 680 million years old. (K/Ar method). The intrusives are not thought to be much younger than the sediments.
Torneträsk, North Sweden	About 600	Fossiliferous strata occur just below a tillite, dated by lithological correlation with more positively-dated equivalents elsewhere in the region.
Podalia, Ukraine, U.S.S.R.	590 *	$^{40}K/^{40}Ar$ methods were used on glauconite, extracted from the sediment.
Rybatschii, Northern Siberia, U.S.S.R.	670 - 900 *	The fossiliferous horizon occurs 100 - 200 m. below a glauconite-bearing layer, dated at 670 million years by $^{40}K/^{40}Ar$ methods.

* Results may be a few percent too high because Soviet scientists use a slightly different half-life for ^{40}K.

To what extent were the various organisms contemporaneous? Table 5.2 summarises some of them, with an estimate of the date of deposition. Some of these are derived from direct or indirect radiometric methods: in other cases this was not possible and dates have been assigned from estimated sedimentation rates, 'filling the gap', from the nearest radiometrically dated horizon.

These, and other results lead us to accept an age of between 590 and 700 million years for the fauna of Ediacara type. A time interval of 110 million years seems very long for the existence of a consistent fauna. It may be as well to remind ourselves that, humanly speaking, even 1 million years is an inconceivably long time! Nevertheless, 110 million years is only equivalent to the time span from the Middle Jurassic to the Upper Cretaceous and there are many genera or even species of the Phanerozoic which existed for this length of time.

The Age of the Earth

We return now to the problem which vexed the early geologists. How old is the earth itself? There are several lines of enquiry which we can follow.

Dating the oldest rocks on earth

Careful searches have been made in Precambrian shield areas to try to find the world's oldest rocks. So far, the Godthab area of south-west Greenland has provided the oldest specimens. These include a conglomerate from Isua containing pebbles of volcanic ash dated by U/Pb methods at 3,824 ± 12 million years, a gneiss from the Amîtsoq Gneisses (about 3,750 million years) and a banded ironstone of about the same age.

The earth must, therefore, be at least 3,800 million years old, but none of these specimens represents the 'original' crust of the earth, since each of them depends for its origin upon the breakdown or metamorphism of yet older rocks. This highlights the major difficulty in trying to find the age of the earth from direct radiometric measurements on material from the earth itself. There have been so many cycles of geological activity: igneous action, metamorphism, weathering and erosion that the original crustal material has long since been recycled several times and the radiometric 'clocks' re-set.

Dating moon rock

It is generally believed that the earth and the moon share a common origin. Whereas the earth's crust, however, is constantly changing under the influence of plate tectonic processes and atmospheric activity, the moon's surface is virtually inert. There is, therefore, a greater chance that primeval rocks may still be found on the surface of the moon and it was one of the aims of the Apollo missions (1969-1972) to bring back representative samples for dating. Many of the specimens proved to be equivalent in age to the earth's oldest rocks, but a significant number of samples gave dates reaching back to 4,600 million years (U/Pb and Rb/Sr methods). The rock types involved are thought to have been original rocks of the lunar crust, excavated later by meteoric impact.

This exciting evidence thus pushes back the origin of the earth-moon system to a minimum of 4,600 million years.

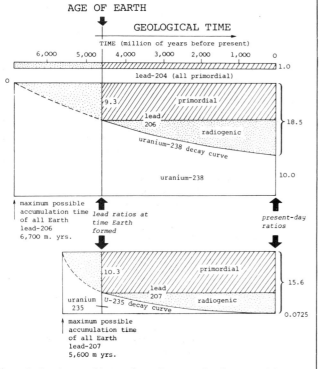

Fig. 5.9 *The radioactive decay of the earth's uranium has added significant lead-206 and lead-207 and has changed their proportions throughout geological time. Left margin of lead-207 curve shows that it cannot have been accumulating for more than 6,700 million years. Indicated ratios are based on lead-204 = 1.0.*

Lead ratios in the earth's crust

Although we cannot find any 'primeval' earth-rock at the surface of the earth, it is possible to derive limiting values for the age of the earth by measurement of the content of lead in crustal material. We have already seen that there are several isotopes of lead, including ^{204}Pb which is non-radiogenic, ^{206}Pb, derived from the decay of ^{238}U and ^{207}Pb, from the decay of ^{235}U. If we assume, for the moment, that <u>all</u> the ^{206}Pb and ^{207}Pb result from the radioactive decay of uranium, there must have been a time when there was no ^{206}Pb and ^{207}Pb, but only uranium. We know the half-lives of the decay series and we can measure the present-day relative abundance of the various isotopes. It should, therefore, be possible to plot back to the time when there was no ^{206}Pb or ^{207}Pb. Modern ratios of the isotopes have been obtained from careful measurements on red clays and manganese nodules from the ocean floors and they are as follows (expressed relative to ^{204}Pb as unity):

^{204}Pb	^{206}Pb	^{207}Pb	^{238}U	^{235}U
1.00	18.5	15.6	10.0	0.0725

Taking these ratios for each of the decay series and plotting back using the known half-lives produces the graphs in Fig. 5.9.

These show that for the $^{238}U/^{206}Pb$ series the maximum accumulation time of all the ^{206}Pb would have been 6,700 million years. The corresponding maximum date from the $^{235}U/^{207}Pb$ is 5,600 million years.

It must be stressed that these are <u>maximum</u> dates. The very discrepancy between them would suggest that the original simple assumption was wrong: in reality there must have been some primeval ^{206}Pb and ^{207}Pb. At least, however, we can say that the age of the earth lies between 5,600 million years and 3,800 million years or so (from the oldest rock). Can we do any better than this?

Lead ratios in meteorites

If we could find out the original proportions of radiogenic and primeval ^{207}Pb and ^{206}Pb, we should be able to arrive at a more precise date for the origin of the earth. The ancient rocks of the earth itself have been affected too much by later processes to be of any value in such measurements. This does not apply, however, to meteorites, which have probably remained unaltered in space, since the formation of the planet from which they fragmented. The meteorite's ancestral planet is assumed to have formed at the same time as the earth, so when meteorites fall to the earth from time to time they provide valuable evidence of what our own planet might have been like.

Measurements have been made on stony meteorites which contain both uranium and lead, and on iron meteorites which contain primeval lead only, but no uranium. The resulting ratios for primeval lead are:

^{204}Pb	^{206}Pb	^{207}Pb
1.00	9.3	10.3

When these ratios are plotted on the graph (Fig. 5.9) both decay series give a 'starting point' about 4,600 million years ago.

Some meteorites can also be dated by direct U/Pb and Rb/Sr methods and the same date of 4,600 million years is obtained.

Summary

Several lines of evidence point to a date of 4,600 million years for meteorites, and possibly for the moon. Earth materials themselves have provided limiting dates for the origin of the earth of between 3,800 million years and 5,600 million years, so an age of 4,600 million years for the earth itself is quite possible. Recently, much detailed work has been done on the lead ratios of lead ore bodies and other rocks, the argument being that when ore bodies are formed the lead is removed from any uranium and the isotopic composition 'frozen'. When the measured lead ratios are plotted on a graph, the 'best fit' time-line (isochron) is one of about 4,600 million years (Figure 5.10). This also passes through the points for meteoric lead showing that the earth and the ancestral bodies from

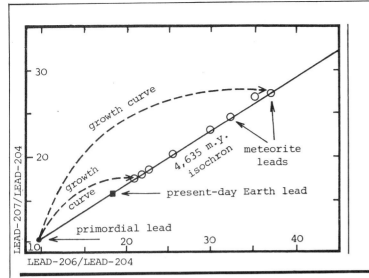

which meteorites are produced have a common origin. This took place about 4,600 million years ago, but geological evidence does not permit us to speculate about exactly <u>what</u> happened then!

Fig. 5.10 Lead isochron diagram for meteorites gives an age of 4,635 million years. Present-day Earth lead falls on the isochron, indicating that it came from the same primordial source as meteorite lead and at the same time.

Continents 19: Oceans 1

We have already seen how the search for the world's oldest rock has so far been rewarded with the discovery of the Isua conglomerate of West Greenland, dating back 3,824 million years. Not so many decades ago, it was predicted by more than one geologist that the world's oldest rocks would be found not on the continents, but on the ocean floors. Such have been the advances in oceanic research in recent years, however, that we now know that the ocean floors are nowhere older than 200 million years; nineteen times younger than the West Greenland rocks! The stark contrast in ages of continental and oceanic crust is shown in Fig. 5.11 where the 'age map' of North America is set against that of the North Atlantic. What part had radioisotopic dating to play in constructing such maps?

The age of the continental crust

The map of North America is not a geological map in the usual sense. It does not show the geological systems as outcrops, but rather, it depicts the date by which different regions had become stable, i.e. were no longer subject to deformation by orogenic activity but only suffered large-scale uplift or depression. It

will be seen that by far the largest proportion of the continent was stable by the start of Cambrian times. Before the advent of radio-isotopic dating, the drawing of such a map could never have been possible, since so much of our dating of the Precambrian depends upon radioisotopic methods.

The pattern of ages revealed in Fig. 5.11 is of considerable interest. It shows that the rocks of the continent are 'zoned' roughly concentrically with the oldest in the middle. At present there are two hypotheses to explain this pattern.

1. The continental crust was all formed by differentiation of the original matter of the earth early in its history (probably in the first thousand million years). Ever since, the same material has been recycled by erosion and sedimentation and re-worked by metamorphic events. Each successive metamorphism has obliterated by 'over-printing' any trace of an earlier date. The concentric arrangement would be explained by the successive collisions and breaking away again of several continental masses under ancient plate tectonic processes.

PATTERN OF AGES IN NORTH AMERICA

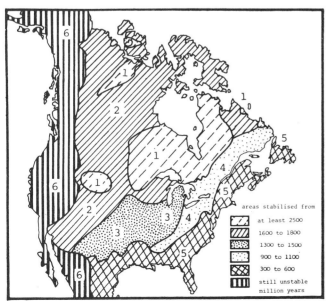

PATTERN OF AGES IN THE NORTH ATLANTIC

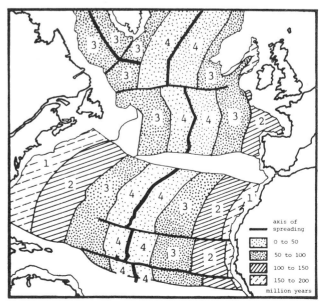

Fig. 5.11   Comparison of ages in North America and in the North Atlantic.

Areas such as north-west Scotland (p.24) are thought to provide supporting evidence for this idea with the Laxfordian gneisses being reworked Scourian material.

2. The continental crust has received additions from time to time by on-going processes of differentiation from the mantle. These processes are non-reversible, with each addition accreting around the margin of the older cores, thus producing the zoned effect.

It is thought that the rocks of West Greenland may provide evidence for this 'model' for crustal growth. In the Godthab region there are two groups of gneisses, the Amîtsoq Gneisses, dated at about 3,750 million years and the Nûk Gneisses, dated at 2,800-3,000 million years ago. At first sight, the situation appears similar to that in North-West Scotland, but it would seem that the Nûk rocks are not simply the younger reworked equivalent of the Amîtsoq ones. Rather, both groups are regarded as having originated by differentiation of the mantle to produce plutonic igneous rocks. In each case, these were metamorphosed within 100 million years or so of their injection to form the respective groups of gneisses.

Those who favour this hypothesis believe that, on a large scale, it could account for the growth of the continental crust. They are not, however, certain of the rate of growth of the crust. It may have been considerably faster in early stages of the earth's history and many geologists believe that some 50-60% of the continental crust was in existence before 2,500 million years ago.

The data on which both these hypotheses are largely based may be thought of as a 'by-product' of radioisotopic dating. As well as measuring the $^{87}Rb/^{87}Sr$ ratios to determine the date, the ratio of radiogenic ^{87}Sr to the non-radiogenic ^{86}Sr is also obtained. Because of the decay of ^{87}Rb, the ratio of $^{87}Sr/^{86}Sr$ must be increasing all the time. By comparing the ratios in crustal rocks with those of meteorites we know that the rate of increase is faster in the crust than in the mantle.

Measuring the ratio of a rock whose age is also being determined, therefore, provides a means of knowing whether it was produced in the mantle or the crust within a few hundred million years of that date.

At present, there is insufficient evidence to show whether or not one hypothesis is a better explanation than the other and opinion is divided.

The age of the oceanic crust

In its own way, the pattern of ages shown in Fig. 5.11 is as characteristic as that of North America. Far from being concentric, however, the pattern is linear and is markedly symmetrical about the Mid-Atlantic Ridge. In this case, research work into the dating of the ocean floors was greatly accelerated by the publication of the sea-floor spreading hypothesis in the 1960s. Indeed, the location of the early survey lines was largely arranged in the most convenient way to test the hypothesis.

The development of the sea-floor spreading hypothesis and the measurement of the ages of sea-bed specimens has been more fully described in the Unit Geophysics. Briefly, there are several approaches to dating the sea floor:

1. Radioisotopic dating of lavas underlying a thin veneer of sea-bed sediment. Specimens are obtained by drilling from a specially adapted ship.
2. Detailed relative dating of sea-bed sediment from its fossil content.
3. Mapping magnetic anomalies and matching alternating lines of reversed and normal polarity to comparable dated sequences on land.

The resulting map (Fig. 5.11) is drawn from data obtained by a combination of these methods and is an elegant demonstration of the validity of the sea-floor spreading hypothesis. Near the ocean ridges the sea floor is less than 10 million years old. Its age then increases with distance from the ridge as the 'conveyor belt' mechanism of sea-floor spreading pushes the earlier-formed material aside. So far no sea bed material has been found with an age much in excess of 200 million years. Far from being permanently in place, as the early geologists believed, the present ocean basins are now seen to be geologically young!

Some have argued that the youth of the ocean floors means that the earth must be expanding, in order to accommodate the new crustal material as it rises at the ocean ridges and the continents are pushed further apart. Most geologists, however, believe that the oceanic crust is 'consumed' by being carried down again towards the mantle in the seismically active areas referred to as destructive plate margins. The complex processes acting at these plate margins would almost certainly reset the 'radioisotopic clocks', quite apart from the fact that the oceanic crust is taken down to the depths where it is completely inaccessible to geologists!

ANSWER TO FIG. 2.7
Oldest to youngest: sandstone, dolerite
quartz-porphyry, limestone

ANSWERS TO FIG. 4.4
a) Between 315 and 331 million years.
b) Less then 315 million years. (You can not
   be more precise in this example.)

Further Reading

Anderton, R., Bridges, P.H., Leeder, M.R. and
Sellwood, B.W.
A Dynamic Stratigraphy of the British Isles.
Allen and Unwin, 1979. A modern text book
which attempts to link the stratigraphy of the
country to plate tectonic processes of the
past.

Donavan, D.T.
Stratigraphy - An Introduction to Principles.
Wiley, 1966. The book lucidly explains the
principles of stratigraphy by reference to
examples, rather than attempting complete
coverage of the geological column.

Eicher, D.L.
Geologic Time. Prentice-Hall, 2nd ed. 1976. A
standard American text introducing the
principles of stratigraphy and geochronology.

Holmes, A.
Holmes Principles of Physical Geology. 3rd ed.
revised by D.L. Holmes, Nelson, 1978.
Professor Holmes was one of the first to apply
radioisotopic measurements to geology and the
book gives a clear account of the methods and
their applications.

Lovell, J.P.B.
The British Isles Through Geological Time.
Allen and Unwin, 1977. An Introductory 'Atlas'
giving the palaeolatitudes and palaeogeographies of the British Isles in very clear
map form, with explanatory text.

Moorbath, S.H.
'The Geological Significance of Early Precambrian Rocks', *Proceedings of the
Geologists' Association*, Vol. 86, part 3,
1975, pp 259-279. This paper is available as
a 'reprint'. It is quite advanced but gives a
very comprehensive account of the early
Precambrian.

Thackray, J.
The Age of the Earth. Geological Museum,
H.M.S.O., 1980. Another in the series of
I.G.S. booklets, profusely illustrated in
colour.

Windley, B.F.
The Evolving Continents. Wiley, 1977.

References

De Geer, G.
Geochronologica Suecica. Principles.
Stockholm, 1940.

Hutton, J.
Theory of the Earth, 1795.

Lyell, C.S.
Principles of Geology, 1830.

Lyell, C.S.
Elements of Geology, 1838.

Playfair, J.
Illustration of Hutton's Theory of the Earth,
1802.

Geophysics

Contents

Acknowledgements

The authors are most grateful to Professor D H Griffiths, Dr R F King and Dr P F Barker of the Department of Earth Sciences, University of Birmingham, for their constructive suggestions and help with sources of information following a reading of the initial manuscript. Much helpful advice has been received from Dr D Livesey, Dr F Spode and Mr J Mansfield. We also wish to acknowledge the willingness of Mobil North Sea Ltd and Esso Petroleum Company Ltd to supply materials from their commercial sources.

The following figures have been based on illustrations from the sources indicated:
Fig. 2.4 and 2.5, *Proceedings of the Yorkshire Geological Society, Vol 34*; Fig. 2.11, Fig. 3.13, Fig. 4.16, Fig. 4.17a, Fig. 4.18, Holmes, A. *Principles of Physical Geology*, Nelson; Fig. 3.5, Dunning, F.W. *Geophysical Exploration*, HMSO; Fig. 3.8, Tarling, D.H. & M.P. *Continental Drift*, Bell and Son; Fig. 3.14 *Understanding the Earth*, Artemis Press; Fig. 4.5 *Esso Magazine No. 116*, Winter 1980/81; Fig. 4.15, Fig. 4.17b, *Unit 22* Open University S100.

1

Introduction

'The earth scientist with the crystal ball is the geophysicist'.
Prof. I G Gass, Open University

'Geologists seem willing to accept a wider range of speculation from the geophysicists than from their own kind.'
Prof. S E Hollingworth, 1957

In a lecture to sixth formers during the 1970s, a well-known speaker showed, to the satisfaction of his audience, that the future of civilised man depended on the work of the geological scientist. The lecturer was Sir Kingsley Dunham, who was then the Director of the Institute of Geological Sciences and most of his audience were studying geology to GCE Advanced Level: a clear case of 'preaching to the converted'! Nonetheless, the talk was a good reminder of how heavily the modern world has come to rely on resources from the ground. 'Fossil' fuels, metal ores, raw materials for the chemical industry, nuclear fuels, building materials – most of these are ultimately derived from geological sources, and all are becoming increasingly difficult to find. The most accessible reserves have long since been worked out and the bulk of current stocks lie buried beneath hundreds or thousands of metres of other rocks, or even beneath the sea bed. This is even more true of the likely discoveries of future reserves.

In such circumstances, surface geological exploration alone is of limited value and increasing use is made of geophysical techniques in prospecting for buried geological structures and materials of possible economic value. As the name implies, geophysics is concerned with the application of the methods of physics to the solution of geological problems. Many of the techniques rely on principles which are first learned in lower-school physics, although the equipment itself has now reached a very high level of sophistication.

In addition to exploration techniques, geophysics has made a major contribution to our knowledge of the deep interior of the earth. Much of this information has been gathered at static geophysical observatories where equipment is used to measure earthquakes and variations in the earth's magnetic field. The relatively young theory of 'plate tectonics' has been formulated largely in order to explain naturally occurring phenomena which have been recorded with geophysical equipment, either at observatories or with mobile instruments.

Table 1.1 lists the main geophysical exploration techniques in current use. Of these, the first three also have important applications to global geology and form the main theme of this Unit.

Used with care, the geological interpretations of geophysical data are invaluable. It will be seen, however, that it is often possible to make several geological interpretations from each set of data and it must not necessarily be assumed that the first interpretation to 'fit' is the only correct one.

Table 1.1

METHOD	MAIN APPLICATIONS
GRAVITY	Land or sea surveys for structures which may contain hydrocarbons. Delimiting sedimentary basins and igneous bodies. Applications of theory of isostasy.
MAGNETIC	Land, sea or airborne surveys, usually at reconnaissance level, to determine depths of 'basement' below possible hydrocarbon fuel traps. Search for ore minerals. Applications to understanding of crustal structure.
SEISMIC a) reflection profile	Determining likely structures for hydrocarbon fuel traps. Measurements of sediment thickness on sea bed.
b) refraction surveys	Site investigation for civil engineering projects. Shallow and deep crustal studies.
ELECTRICAL e.g. resistivity	Ground-based surveys for: site investigation; superficial deposit studies; studies of water table depth.
ELECTRO- MAGNETIC	Mostly airborne surveys for ore mineral deposits.
RADIOMETRIC	Search for radioactive minerals such as uranium and thorium ores. Reconnaissance surveys from the air, followed by ground work.
BOREHOLE LOGGING – a wide range of measurements from probes lowered into boreholes	Obtaining information of rock-type, stratigraphy and structure during the course of borehole exploration for oil, gas, coal, water etc. and for academic purposes. Particularly valuable when boreholes are not being cored, for reasons of expense or time.

2

The Gravity Method

General Principles

Most people are familiar with the concept of gravity, irrespective of the truth or otherwise of the story of the apple falling onto Sir Isaac Newton's head! We are all aware of the problems of weightlessness suffered by astronauts in space and many students have tried to measure the earth's gravitational field in the school physics laboratory, by swinging a simple pendulum or by falling-weight techniques. Results of around 9.8 m s^{-2} may be obtained by such experiments, with the accuracy of a few per cent. This is quite impressive for a school experiment but is of little value for exploration purposes, where measurements must be accurate to at least one part in 10^7!

The units given above indicate that we actually measure the acceleration due to the gravity of the earth rather than the <u>force</u> acting between the earth and the measuring instrument. It is not therefore strictly correct to refer to the earth's gravitational <u>attraction</u>, although we often do!

Absolute and relative gravity

'Absolute' gravity is measured by a more elaborate version of the schools' falling-weight equipment, but since the apparatus is cumbersome, measurements are normally restricted to widely spaced centres, e.g. at geophysical observatories. For most gravity survey purposes, knowledge of absolute gravity is less important than being able to measure accurately and speedily the minute <u>variations</u> in gravity from place to place. For this purpose, a variety of gravity meters (or gravimeters) has been developed, most of which work on the principle of a very delicate spring balance, where an increase in the gravitational field results in greater extension of the spring. Absolute gravity values may readily be assigned to gravity meter stations by 'tying-in' to observatories where such determinations have already been made. In the case of the British Isles the main base is at Cambridge.

Units

Since surveyors are looking for very minor changes in the acceleration due to gravity, the usual S.I. unit is far too large and the unit used in geophysics is the <u>gravity unit</u> (g.u.). One gravity unit is equal to 10^{-6} m s^{-2}. (Older literature and commercial geophysicists use the milligal. One milligal equals 10^{-5} m s^{-2}.)

The earth's gravity field

What are the factors which influence the earth's gravity field? By far the most important is the overall attraction exerted by the great mass of the earth, but superimposed on this are other influences, of which four are of relevance to the exploration geophysicist.

The effect of the underlying rocks

The aim of most gravity surveys is to determine the disposition of the various rock types lying at shallow depth beneath the area of the survey. Since the gravity meter is affected by the mass underneath it, if the rocks are unusually heavy or light, then the reading will vary accordingly when compared with readings for surrounding rocks. One of the earliest applications for gravity surveys was in the search for salt domes in the Gulf Coast oilfields of the USA. These cylindrical masses often act as oil traps and are of much lower density than the surrounding rocks. There is therefore a mass deficiency and a 'negative anomaly' is recorded, i.e. the gravity values are lower than those measured on the adjacent rocks.

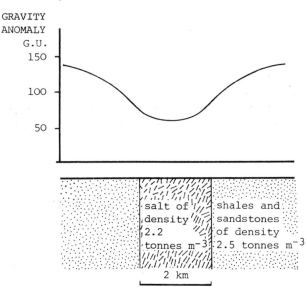

Fig. 2.1 *Simplified gravity profile across a salt dome.*

Further examples of the applications of gravity surveys will appear later in the Unit. Unfortunately, the effects of the sub-surface geology are mostly small compared with the other three influences and corrections have to be made before the survey results are geologically useful.

The latitude effect

It is well known that the earth is not a perfect sphere but more like a rather flattened spheroid. The diameter at the Poles is about 12,714 km in comparison to 12,757 km at the Equator. This means that the surface of the earth at the North and South Poles is nearer to the dense core than is the surface at the Equator. Gravity values at the poles are therefore proportionally higher than those at the Equator. The latitude effect is as great as 15 g.u. per minute of latitude, so very accurate maps are necessary to allow for it in correcting the 'raw' gravity meter readings.

The height effect

The higher one goes above sea level, the further one is from the main mass of the earth and so the gravity readings become less. A modern meter is so sensitive that the difference in gravity between the floor and table height is measurable ($3 \cdot 1$ g.u. m^{-1}) so accurately surveyed heights are essential. Unlike the floor-to-table height analogy, in reality, there is rock material between the gravity station and the datum level used for the survey (usually sea level). This has its own attraction which has to be allowed for, using an assumed or measured density for the rock mass. The correction is known as the Bouguer correction and the corrected gravity figures in most common use are referred to as the Bouguer anomaly (Fig. 2.2).

The terrain effect

Variations in local relief also exert an influence on the tiny mass in the moving system of the gravity meter. A further correction has therefore to be added. It is calculated using a special grid superimposed on the relief map of the surveyed area.

In very accurate work, the attraction of the sun and moon must also be computed.

Principles of interpretation

The need for data processing described above makes the conduct of a gravity survey sound very laborious. Much of it is rather routine, although computers are now used to speed the calculation procedure. The excitement lies in the interpretation of the Bouguer anomalies!

Before we proceed with some examples, it must be appreciated that the interpretations which usually appear in publications are the best selected from a variety of possibilities. In theory, a large number of underground distributions of rock types might explain a measured Bouguer anomaly. In practice an estimate of what might be happening below ground is drawn up as a 'model'. This is done on the basis of the known surface geology of the area, of measurements of the density of the rocks exposed and on experience of similar geological situations elsewhere. If borehole information is available it is, of course, particularly useful.

The gravity anomaly which would be produced by the 'model' is computed from standard formulae and compared with the measured one. Adjustments are made to the 'model' where there is disagreement and the computation repeated. This may need doing many times until the measured anomaly is matched sufficiently accurately for the purpose of the survey.

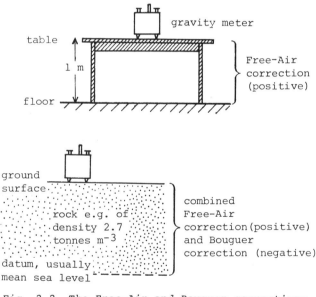

Fig. 2.2 The Free-Air and Bouguer corrections.

Some Case Studies of Interpretations of Gravity Data

An Antarctic glacier

Gravity surveys have been used to estimate the thickness of ice in glaciers for geomorphological purposes, and to add to the accuracy of our knowledge of the total quantity of ice in the world. The example given in Fig. 2.3 is a single gravity profile across the Starbuck glacier in the Antarctic Peninsula (lat. 65° 38' S) beginning and ending on the rock sides. Here a huge density contrast exists between the glacier ice and the rock walls and bed of the valley. Laboratory measurements of the density of glacier ice and of specimens of bedrock (in

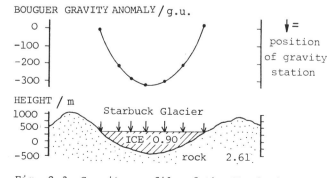

Fig. 2.3 Gravity profile of the Starbuck Glacier. (Vertical and horizontal scales are equal). Densities in tonnes m^{-3}.

this case a granite) gave figures of 0.9 tonnes m^{-3} and 2.61 tonnes m^{-3} respectively, a density contrast of about 1.7 tonnes m^{-3}.

Although there may be density changes within the bedrock below the glacier, they will be small in comparison to the large contrast between the ice and the granite and are unlikely to alter the interpretation very much. The diagram shows a good fit between the measured anomaly shown with a line and that computed for the 'model' shown by points. Evidently there are still some 700 m of ice in the valley, with a further 700 m of towering rock wall above the ice surface.

The Stranraer sedimentary basin

A major application for the gravity method is in delimiting sedimentary basins as an early step in the exploration for oil reserves although the example shown in Fig. 2.4 was surveyed for academic purposes rather than economic ones. It is the Stranraer sedimentary basin in the Southern Uplands of Scotland and it consists largely of Permo-Triassic sandstones and breccias. These lie at right angles to the strike of the dominant Lower Palaeozoic rocks which comprise the rest of the Southern Uplands. A borehole for water had shown that the Permo-Triassic rocks were at least 190 m thick and it was suspected that they were much thicker than this.

The gravity survey revealed a Bouguer anomaly of over 130 g.u. below the background values on the Lower Palaeozoic rocks. Figure 2.5 shows the measured anomaly along the line of section CC', with, beneath it, three possible interpretations of the cross-section of the Permo-Triassic basin. This has been general-

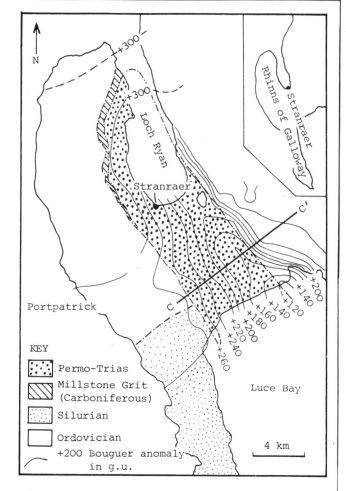

Fig. 2.4 *The geology and Bouguer anomalies of the Stranraer district.*

ised into a series of steps for ease of computation. A density contrast of 0.4 tonnes m^{-3} (model 2) was considered most probable, from surface measurements, but it is geologi-

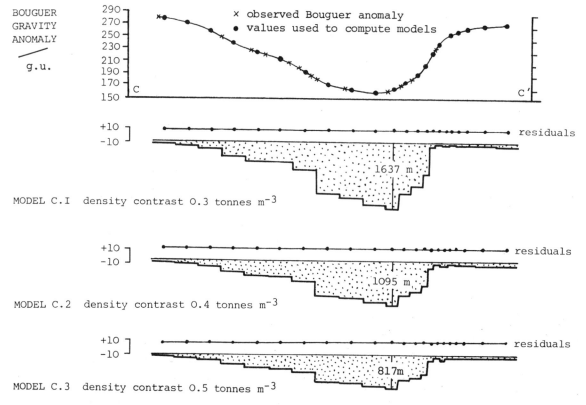

Fig. 2.5 *Postulated shapes of the Stranraer sedimentary basin along the line CC' (see Fig. 2.4) for a range of density contrasts.*

cally possible that the contrast could be as low as 0.3 tonnes m^{-3} or as high as 0.5 tonnes m^{-3}. The corresponding depths for these density contrasts are shown as models 1 and 3. Even with the uncertainty about the density contrast, the basin proved to be many times deeper than had been thought previously. Furthermore, the gravity gradient was so steep on the eastern side of the basin that it appears almost certain to be due to a normal fault.

Similarly shaped negative gravity anomalies are also common over granites, whose density is usually less than the surrounding country rock, so one must not immediately assume that every low gravity anomaly represents a sedimentary basin.

The Insch Gabbro, Grampian

Before you read on, try to predict the shape of the anomaly which you would expect over a gabbro body, given that gabbro is largely composed of plagioclase (density 2.71 tonnes m^{-3}) and augite (3.3 tonnes m^{-3}) plus possibly olivine (3.7 tonnes m^{3}) whilst granite is mostly orthoclase (2.57 tonnes m^{-3}), plagioclase, quartz (2.65 tonnes m^{3}) and micas (about 2.9 tonnes m^{-3}).

The section in Fig. 2.6 is a north-south gravity profile of the Insch Gabbro and the adjoining Bennachie Granite in Grampian, Scotland. Not suprisingly, the anomaly is higher over the gabbro. A computed 'model' which provides a reasonable fit is given beneath the gravity profile, and shows the gabbro to be a sheet-like intrusion.

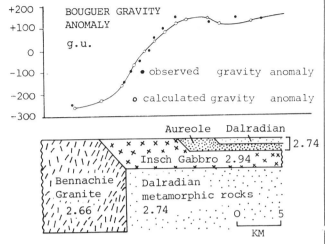

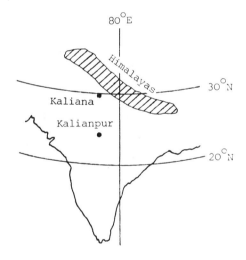

Fig. 2.6 Profile of gravity anomalies across the Insch Gabbro compared with a theoretical profile calculated for the model below. Densities are tonnes m^{-3}.

Global Applications of the Gravity Method

Long before gravity meters had been developed, surveyors had become aware of local irregularities in the earth's gravity field. Indeed, the term 'Bouguer anomaly' is named after a French surveyor, Pierre Bouguer, who worked in the Andes in the eighteenth century. It was not until the middle of the nineteenth century, however, that the implications of changes in the gravity field were fully appreciated and attempts made to match surface measurements with computed variations in the earth's crust. The hypotheses which emerged affect our understanding of major crustal structures today, so a brief account of how they arose is not out of place here.

In the 1840s and 1850s, surveyors were busy mapping accurately the land surface of the Indian sub-continent, mainly to aid the British government in its administration of this huge territory. Two methods were in use, the one complementing the other. Astronomical 'fixes' were painstakingly made at various survey stations, which were then linked by overland measurements using triangulation. In the normal course of events, the results from the two different methods should agree. However, on traverses running north to south, which came within reach of the Himalayan mountain range, considerable errors were noted. For example, the difference in latitude between two stations, Kaliana and Kalianpur (Fig. 2.7) were measured as follows:

by triangulation: 5° 23′ 42″.29
by astronomical fixes: 5° 23′ 37″.06

Fig. 2.7 Location of Kaliana and Kalianpur in relation to the Himalayas.

This shows a discrepancy of 5.23 seconds of arc, which does not sound very much, but it represents an error of over 150 metres on the ground. The surveyors knew that their instruments were more reliable than these figures suggested and were not slow to seek an explanation.

The method of position fixing by astronomical means demands a very accurate knowledge of the vertical, whilst triangulation is less dependent on this knowledge. It was therefore suggested that perhaps the Himalayas were exerting a significant sideways pull on the levelling mechanism of the instrument for the Kaliana astro-fix, but that Kalianpur was too distant for it to be so badly affected. This is shown in highly exaggerated form in Fig. 2.8, positions a and b.

a = true vertical
b = actual deflection of plumb bob
c = calculated deflection of plumb bob

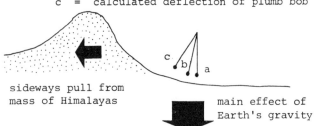

sideways pull from mass of Himalayas

main effect of Earth's gravity

Fig. 2.8 *The actual and supposed gravitational effect of the Himalayas on the survey instruments.*

Calculations of the probable mass of the Himalayas were made, from their known volume and the density of surface rock samples, but far from matching the observed discrepancy, the calculated deflection produced by the sideways pull of this mass (c in diagram) was three times as great as the observed deflection (b). Why was this?

Up to that time, mountain ranges had been regarded as additional masses 'stuck' on to an otherwise rigid crust of the earth. The density of the crust was assumed to be much the same from place to place. Perhaps these old assumptions were wrong!

wood blocks are of different densities

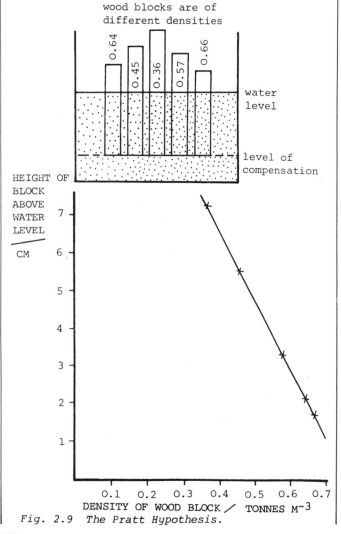

HEIGHT OF BLOCK ABOVE WATER LEVEL

CM

water level

level of compensation

7
6
5
4
3
2
1

0.1 0.2 0.3 0.4 0.5 0.6 0.7
DENSITY OF WOOD BLOCK / TONNES M⁻³

Fig. 2.9 *The Pratt Hypothesis.*

In 1855, two rival hypotheses were published in explanation, each of which proposed that a state of hydrostatic balance existed in the uppermost layers of the earth. Thus J H Pratt suggested that the mountain ranges had risen like fermenting dough - the higher they rose, the less was the density of the rocks composing them. In simple terms, Pratt's hypothesis may be illustrated by floating blocks of wood of varying density in water. The highest blocks are of lowest density and vice-versa. A model to test this idea may be made fairly easily, with the wood blocks free to slide up and down supporting wires to prevent them from toppling over (Fig. 2.9) Note that the bases of the floating blocks lie at an even 'level of compensation', which accorded with current thinking about the crustal layers in the mid-nineteenth century. Beside the diagram is a graph drawn from observations on a home-made model of the height of each block out of the water plotted against its density. *Does the shape of the graph suggest that Pratt's hypothesis is one possible explanation?*

Wood blocks are of the same density (0.62)

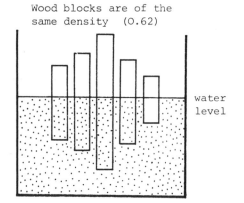

water level

HEIGHT OF BLOCK ABOVE WATER LEVEL

CM

6
5
4
3
2
1

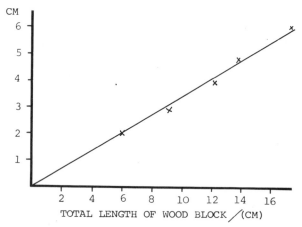

2   4   6   8   10   12   14   16
TOTAL LENGTH OF WOOD BLOCK /(CM)

Fig. 2.10 *The Airy Hypothesis.*

The alternative hypothesis, of G B Airy, proposed that the mountain ranges were behaving rather like icebergs, with a deep 'root' beneath the highest ranges, penetrating into the denser substratum. This idea, too, may be illustrated with wood blocks in water, only here the blocks are all of the same density (Fig. 2.10). Compensation for height is achieved by the block floating more deeply in the water. In this case, the graph shows the height of each block out of the water plotted against its total length. *Does the graph suggest that this hypothesis is also viable?*

Isostasy

Each man had offered an explanation that a state of hydrostatic balance could exist in the earth's surface layers, even though the 'substratum' is not liquid, but rather acts as a plastic substance over long periods of time. Such a state of balance later became known as <u>isostasy</u> (i.e. 'equal stability'), but it was many years before other methods became viable to help test the two hypotheses. Seismic work (described in a later section) has shown that beneath the continents the crust is considerably thicker and of lower density than it is beneath the oceans. To this extent, Airy's hypothesis seems to have the more widespread validity. However, recent seismic work demonstrates that there are <u>some</u> regions, particularly certain mountain ranges, where lateral changes in density offer a better explanation than a deep 'root' of low density material. Thus, Pratt's ideas seem to be vindicated too, albeit on a smaller scale.

Subsequent geophysical work has shown that such a state of isostatic balance does indeed exist throughout much of the world, but there are some exceptional areas where there are huge gravity anomalies, representing considerable mass deficiencies or surpluses and therefore a lack of isostatic balance. Presumably, in such areas, geological processes have affected the crust and upper mantle faster than isostatic adjustments can effect a re-balancing. One area of mass deficiency is North Sweden. This region was one of the last parts of Europe to be deglaciated over the last few thousand years in the wake of the last Ice Age. It would appear that the huge weight of the ice sheet depressed the light rocks of the crust, causing displacement of mantle rocks beneath. The ice then melted very rapidly, in geological terms, and there has not yet been sufficient time for the balance to be re-adjusted by 'flow' within the mantle. There

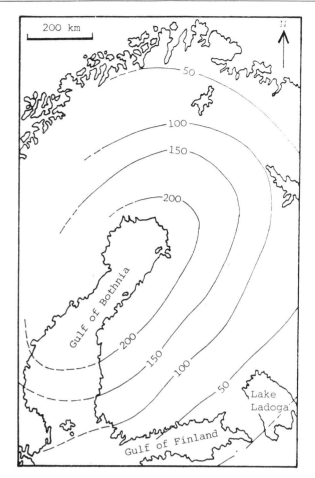

Fig. 2.11 Post-glacial uplift of Fenno-scandia (Finland and Scandinavia). The curves are lines of equal uplift, in metres, from 6800 BC to the present day. Around the northern shores of the Gulf of Bothnia the present rate of uplift is 1 cm per year.

is therefore a mass deficiency beneath North Sweden which is reflected in the negative gravity anomaly and the region is still rising at a measurable rate relative to sea level (Fig. 2.11).

3

The Magnetic Method

General Principles

Most people have used a magnetic compass and know that the magnetised compass needle points north-south. It is also common knowledge that the needle does not point to the geographic north and south poles but rather is aligned towards the magnetic poles, which are several hundred kilometres away from the geographic poles. The exact position of each magnetic pole changes slowly over the years as the magnetic pole 'precesses' around the geographic one. At present in the English Midlands, magnetic north lies some 7° west of true north, decreasing annually. This angle is called the declination (Fig. 3.1).

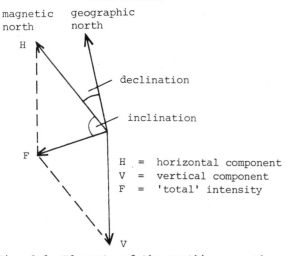

Fig. 3.1 Elements of the earth's magnetic field.

Near the poles themselves, the horizontally-acting component of the earth's field, which has most influence on a compass needle, is very weak. The needle is also being pulled downwards and cannot swing freely, so it becomes very sluggish. In fact, the magnetic pole itself was located by taking measurements with a dip circle, where a magnetised needle is suspended on a horizontal axis. The pole lies beneath the point where the needle reads 90° from the horizontal. On the magnetic equator the dip circle reads zero (Fig. 3.2). The angle from the horizontal of the magnetic dip is known as the inclination (Fig. 3.1).

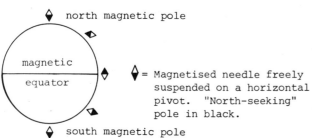

Fig. 3.2 Magnetic latitude.

These well-known instruments are simply responding to the magnetic field of the earth. Experiments are often conducted in the lower school with iron filings or by plotting-compasses around a laboratory bar magnet. Figure 3.3. shows the usual sort of two-dimensional pattern which is produced by iron filings when they are sprinkled on a sheet of card placed over the magnet. The field would be similar if it could be seen in the third dimension too.

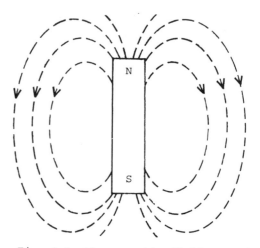

Fig. 3.3 The magnetic field around a bar magnet.

Although no one pretends that the earth has a bar magnet inside it, the magnetic field of the earth is remarkably similar to the one just described. Certainly it is di-polar (i.e. it has a north and south pole) and has apparently been so for most of the earth's history. It probably owes its origin to an internal dynamo effect, produced by thermal convection in the electrically conducting liquid core.

Magnetic field strength

The strength of the earth's magnetic field varies, for three main reasons. First, internal motion in the core produces slow secular changes in field strength, measurable over a period of years. Secondly, there are diurnal changes, brought about by atmospheric disturbances on a day to day basis. Diurnal effects are at their most extreme at times of increased sun-spot activity. Thirdly, and of most interest to exploration geophysicists, variations in near-surface rock types may produce measurable local differences, or anomalies in the magnetic field.

A variety of instruments, known as magnetometers, has been developed to measure such variations in the magnetic field. Earlier instruments mostly depended on delicately pivoted needles, but modern surveyors usually use electronic equipment which is readily adapted for surveys on the ground, in the air or at sea.

Units

The unit used in magnetic work is the nano-tesla (nT), based on the S.I. unit, the Tesla. In older surveys the term gamma was used but this is numerically the same as the nT so conversion is easy. The field strength of the earth ranges from about 25,000 nT near the magnetic equator to about 70,000 nT near the magnetic poles.

Some branches of geophysics are concerned with the overall pattern of the earth's magnetic field and maps are available which show the field values and indicate the slow secular changes over the years. Like the earth's gravity, however, there are minor irregularities, or anomalies, superimposed on the main field. Geologically significant anomalies may be up to several thousand nT in amplitude.

Why are there such variations in the magnetic field? There are two main reasons of relevance to the geophysicist.

Induced magnetisation

Many rocks contain iron-bearing minerals, such as magnetite, which are capable of having a magnetisation induced in them by the action of the earth's field today. The ease with which a rock may be magnetised is known as its susceptibility and rocks have characteristic values which may be measured in the laboratory. The susceptibility of most sedimentary rocks is low (but seldom completely negligible). Acid igneous rocks and most metamorphic rocks are about ten times more susceptible, whilst most basic igneous rocks are about one hundred times more susceptible than the sedimentary ones.

Here then, is a potentially useful tool for mapping the sub-surface distribution of crystalline 'basement' rocks, basic igneous bodies like dykes and sills and magnetic ore bodies.

Remanent magnetisation

A second factor, however, must also be considered. This is, the rock may already be

'magnetic' in its own right, resulting from a previous phase in its history, and not connected with its position in the earth's present magnetic field. Such a property is called the remanent magnetisation and it may even be stronger than the induced magnetisation of today. It may also lie in a completely different direction.

The main way in which a rock may acquire such a remanent magnetisation is when it cools from the molten state. When molten, it cannot retain a magnetisation, but once crystallised, it eventually passes a certain critical temperature, known as the Curie Point, below which it can. It then becomes magnetised in the prevailing direction of the earth's field at the time.

Remanent magnetisation may be produced in sedimentary rocks when detrital grains of magnetic minerals settle from suspension. Also, it is common for magnetisation to be produced during the chemical changes which accompany the oxidation of red beds.

Generally, the remanent magnetisation acquired by sedimentary rocks is small in comparison to that in igneous rocks, but there are some circumstances where it is of great interest.

The magnetic anomaly measured at the earth's surface really depends, therefore, on the resultant of two types of magnetisation – induced and remanent (Fig. 3.4).

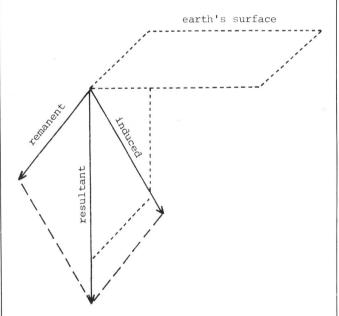

Fig. 3.4 Resultant magnetisation.

Magnetic Survey Work in Areas of Continental Crust

A local survey across an ore body

One of the simplest applications of the magnetic method is in the search for dyke-like bodies of igneous rocks or injections of magnetic ore minerals in otherwise 'non-magnetic' terrain. The example in Fig. 3.5 is from northern Ontario, Canada. A single

magnetometer profile across the magnetite veins is shown, together with one possible interpretation of the sub-surface geology. In this case, it is assumed that the anomaly results from the induced magnetisation of the ore body: if its remanent magnetisation is to be considered too, several other inter-pretations are possible.

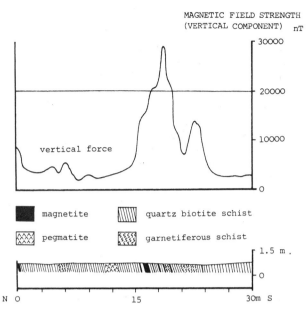

MAGNETIC FIELD STRENGTH
(VERTICAL COMPONENT)   nT

Fig. 3.5 Profile of vertical component of magnetic field across magnetite iron ore layers near Nakina, Northern Ontario, from ground survey.

An aeromagnetic survey

When an oil company begins to explore a new area for hydrocarbon fuels it usually commissions an aeromagnetic survey early in the exploration programme. The method is relatively cheap and fast, and when the aircraft is flying over land it may easily be combined with a programme of aerial photography, for use in structural geological mapping.

Much of the North Sea was covered by airborne magnetometer surveys before further geo-physical work was carried out. Figure 3.6 is an extract from an aeromagnetic map and it covers the part of the North Sea where the Beryl Oil Field was subsequently discovered. The location of Block 9/13, which contains the Beryl Field is given in Fig. 4.2.

Figure 3.6 shows that there is a positive anomaly within the block, indicating that magnetic 'basement' rocks are nearer to the surface here than in the immediate surroundings. The steepness of the magnetic gradients suggests that block-faulting is responsible for bringing up the basement rocks and the likely positions of these faults are shown by the double lines in Fig. 3.6.

The axes of the positive anomalies are shown by dotted lines, and depths to magnetic basement have been calculated at points along these axes. Thus, for Block 9/13 the calculated depth is 14,600 feet (4450 metres). Later in the development of this field, holes were drilled, but none of them has yet penetrated the magnetic basement, so its nature in this part of the North Sea is unknown.

The magnetic map itself does not prove the presence of oil, but it does suggest areas where the crystalline 'basement' rocks of the continental crust lie nearer to the surface than usual. These 'basement highs' may relate to structures in the overlying sedimentary rocks which might be favourable for the trapping of oil or natural gas. Again, remanent magnetisation is usually ignored in the commercial companies' inter-pretations of the magnetic data. The impor-tance of this particular example is pursued in Section 4.

Magnetic Studies for Global Geological Purposes

In the preceding applications of magnetic work, we have mostly considered the induced magnetisation to be the dominant factor in producing a magnetic anomaly. However, since the late 1950s there have been many signi-ficant advances in our understanding of the earth, due in no small measure to considera-tion of the importance of remanent magnetisation. This section will study the measurement of remanence, its application to theories of the history of the continents and lastly its contribution to the magnetic anomalies of the ocean floors. Since remanent magnetisation was produced in the geological past, its study is normally referred to as palaeomagnetism.

Palaeomagnetism and continental drift

Laboratory work is conducted to determine the remanent magnetisation of specimens of 'magnetic' rocks brought back from the field. A portable, petrol-driven rock drill is used to extract small cylinders of rock from the exposure. They are marked with a north arrow and the angle of the specimen from the horizontal is noted, so that it may be re-oriented in the laboratory. The tectonic dip is also noted, so that the effect on the direction of magnetisation of subsequent earth movements may be corrected.

Once in the laboratory, the direction and strength of the remanent magnetisation in the sample is measured by a specially developed magnetometer and an elaborate series of further tests is done to check the reliability of the data. Average values are computed from the results of measurements of several specimens from the same site. After any necessary corrections, the result gives the direction of the earth's field <u>at the time when the rock was magnetised.</u>

Igneous rocks are usually regarded as the most fruitful rock type for such studies, but modern equipment is now delicate enough to be able to cope with the more weakly magnetised sedimentary rocks indicated earlier.

As this kind of research developed in the late 1950s, some rather surprising results began to emerge which have received

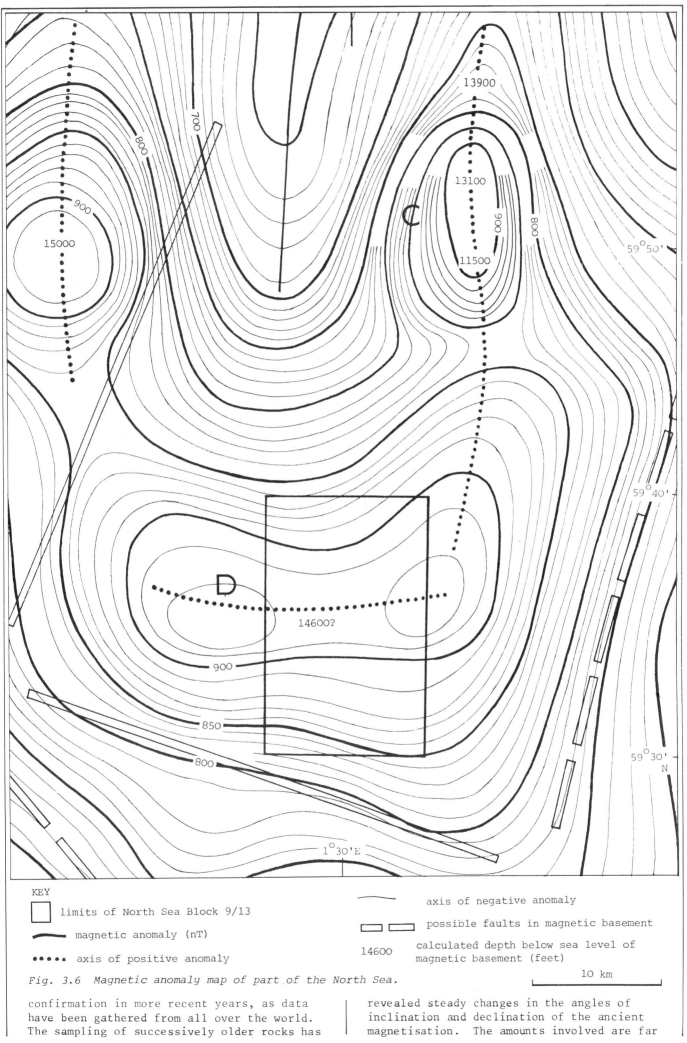

KEY

☐ limits of North Sea Block 9/13

▬▬ magnetic anomaly (nT)

••••• axis of positive anomaly

⌒ axis of negative anomaly

▭ ▭ possible faults in magnetic basement

14600 calculated depth below sea level of magnetic basement (feet)

10 km

Fig. 3.6 Magnetic anomaly map of part of the North Sea.

confirmation in more recent years, as data have been gathered from all over the world. The sampling of successively older rocks has

revealed steady changes in the angles of inclination and declination of the ancient magnetisation. The amounts involved are far

Table 3.1

GEOLOGICAL PERIOD	DATE IN MILLION YEARS	PALAEOMAGNETIC INCLINATION (I)	PALAEO-LATITUDE (L)
Present	0	69°	
Tertiary	50	61°	
Cretaceous	100	57°	
Jurassic	170	50°	
Triassic	220	50°	
Permian	250	23°	
Carboniferous	300	0°	

more than can be explained by the relatively slight precession of the geomagnetic poles around the geographical ones. Indeed, it is generally believed that over long periods of time, the magnetic and geographical poles have had the same average position.

The significance of the changes in direction may be seen from measurements on many British rocks. During the Carboniferous Period, the palaeomagnetic inclination was about 0° from the horizontal, so the British Isles must then have been on the Equator. Working upwards through the succession, the value steadily increased until today, the magnetic dip is about 69°.

Assuming that the earth's field has been dipolar, at least since the Carboniferous, it is possible to plot the changing position of Britain in relation to the Poles. Table 3.1 gives some generalised palaeomagnetic inclinations for Britain at various times since the Carboniferous.

Fortunately there is a simple relationship between the angle of inclination of the remanent magnetisation and the ancient, or palaeo-latitude, expressed in the equation:

$$\text{Tan } I = 2 \text{ Tan } L, \text{ where}$$
$$I = \text{angle of inclination}$$
$$L = \text{latitude (Fig. 3.7)}$$

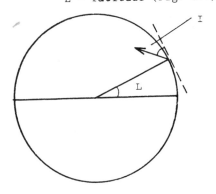

Fig. 3.7 Relationship between latitude (L) and magnetic inclination, or dip (I).

Using this equation, calculate the changing palaeolatitude of Britain and plot the positions on a globe or on an Atlas projection. Note that it is not possible to find the palaeolongitude by this method.

In carrying out this plot, you may have satisfied yourself that you have 'proved' that a portion of the earth's crust may move about on the earth's surface in relation to

the poles. On a wider scale, this has become known as the 'Continental Drift Hypothesis'. Have you proved this, however? What if Britain had remained stationary and it was the earth's magnetic axis which had been moving? In the 1950s, the hypothesis of continental drift was not favoured by European and American geologists and the alternative idea, known as the 'Polar Wandering Hypothesis', was given strong consideration. Indeed, you could equally well redraw your map, showing Great Britain at the same position on the globe and a series of points to represent the successive positions of the poles. Joining these points would produce an 'apparent polar wandering curve'.

Later, however, it was shown that the magnetic poles have probably never 'wandered' more than a few hundred miles from the geographic poles and that the latter are unlikely to have moved to anything like the extent required by the polar wandering hypothesis.

As palaeomagnetic results began to be derived from other continents too, the continental drift explanation seemed to be the more plausible. Workers were each plotting their apparent 'polar wandering curves' for their own particular continent, but these frequently disagreed, and only converged onto a single North or South Pole for the present day! Were there, then, dozens of different poles in the past? In spite of the sceptics the simpler solution seemed to be to 'slide' continents about the globe until their 'apparent polar wandering curves' coincided for large periods of geological time (Fig. 3.8). The resultant pattern was remarkably similar to the various attempts to fit the continents together on the basis of the edges of the continental structure at 1000 m below sea level. Further checks were made to see how well the continental reconstructions on the basis of palaeomagnetic latitudes tallied with the ancient climates, distribution of fossils and structural connections known already from geological work. Again the match was remarkably close. Geologists who had worked in the Southern Hemisphere, believing all along in continental drift, now had every reason to be smug, whilst their colleagues in the Northern Hemisphere had the grace to metaphorically eat their earlier words and agree that if everything fitted in so well, then continental drift must, after all, be a 'good thing'!

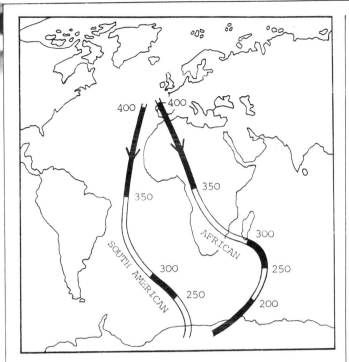

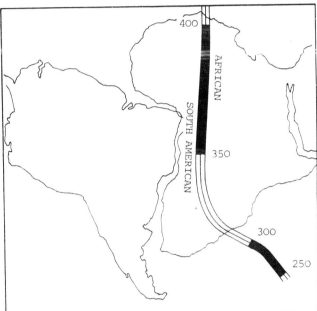

Fig. 3.8 the fit of Africa and South America to reconcile their apparent polar wandering curves: a) The apparent polar wandering curves for the two continents in their present position (dates in millions of years) b) Moving the continents to match their apparent polar wandering curves produces a fit identical to that required by geological observations for at least 150 million years.

Reversals of the earth's magnetic field

We have become accustomed to the north-seeking end of a compass needle always pointing to the North Pole and we would be somewhat surprised if it suddenly pointed South! As the number of palaeomagnetic measurements grew, however, it became clear that such a reversal of polarity had happened, not just once in the past, but many times.

In rocks such as lavas it is also possible to date the different periods of normal or reversed magnetisation by radiometric means (see the Unit *Geochronology*). At first, lavas

were collected for this purpose from successive flows in areas such as Iceland and a chronology worked out for the last 4½ million years or so. Beyond that, the radiometric method was not accurate enough. However, once techniques for coring at sea were developed it became possible to collect fossil-bearing sediments immediately over-lying the lavas. This has recently enabled the production of the time scale shown in Fig. 3.9, which can accurately be extended back to the Upper Cretaceous, some 70 million years ago.

Many other periods of reversed polarity have also been recognised throughout older parts of the geological time scale.

From the geological record, reversals appear to happen suddenly, but it is probable that the magnetic field dies away to nothing and then builds up again in the opposite direction over a time span of several thousand years.

At first, reversals were regarded as a scientific curiosity, but they later became of great importance in aiding our under-standing of the geology of the ocean floors, as the next section shows.

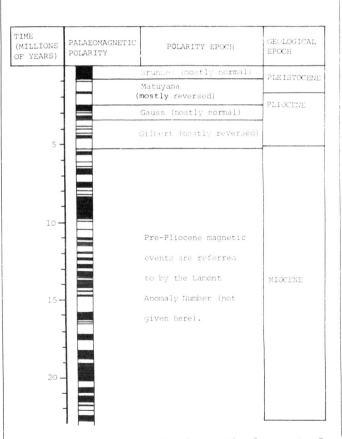

Fig. 3.9 The magnetic time scale for part of the Tertiary and Quaternary. Shaded blocks represent normal polarity; open blocks represent reversed polarity

Magnetic survey work in the ocean basins

During the late 1950s and early 1960s a considerable number of research ships were busy surveying the ocean floors with echo-sounders to determine their relief features. At the same time, each ship could also tow the detector unit of a magnetometer, originally developed as a rather primitive means of submarine spotting! Detailed maps of the type shown in Fig. 3.10 were produced. Rather than showing magnetic contours, as most previous aeromagnetic maps had done, the map depicts positive magnetic anomalies in black and negative ones in white.

The magnetic anomalies revealed a far more complex pattern than many geologists had expected. However, apart from picking out obvious fracture zones in the rocks below the sea bed, no really satisfactory interpretation could be made at first. No one could be sure how much of the anomaly was due to susceptibility contrasts and how much was the result of changes in remanent magnetisation. The rocks of the deep ocean floor were then quite inaccessible, so oriented specimens could not be brought back for laboratory measurements of remanence.

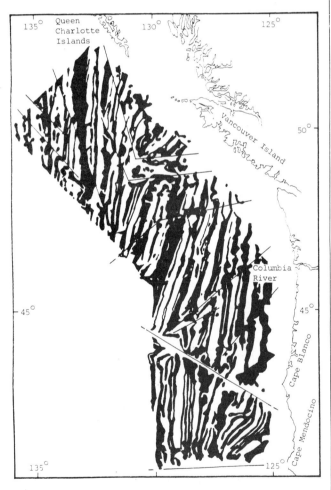

Fig. 3.10 *Magnetic anomaly patterns in the north-east Pacific off the Canadian and United States coasts. Straight lines indicate faults displacing the anomaly pattern.*

The Vine and Matthews hypothesis

In 1963, two British geophysicists, F J Vine and D H Matthews published a very short paper suggesting a clue to the problem. They pointed out that if the rocks producing the

anomalies were of igneous origin, the remanent magnetisation could well be of greater importance than any susceptibility contrast. If the sea bed were underlain by adjacent intrusions or lava flows produced at intervals over several million years, it was quite probable that some would have been formed at times when the earth's magnetic field was normal, and others when it was reversed. Igneous bodies of normal polarity would 'reinforce' the modern earth's field, producing a positive anomaly; bodies of reversed polarity would 'subtract' from it and give rise to a negative anomaly (Fig. 3.11).

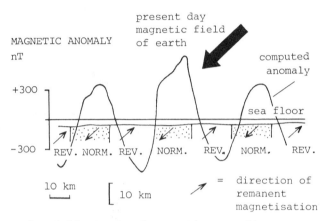

Fig. 3.11 *Computed magnetic anomalies across a series of reversed and normally magnetised blocks of oceanic crust.*

Supporting evidence for their ideas came from an abortive attempt to drill a hole right through the earth's crust into the mantle, which took place at about this time. Drill cores were recovered which showed reversally magnetised basalts lying at shallow depth below the sea floor.

Sea-floor spreading

The previous year, an American geologist, H H Hess, had attempted to explain the origin of the oceanic ridges, such as the Mid-Atlantic Ridge. He had suggested that beneath the ridges the rocks of the upper mantle were being partly melted, releasing basaltic magmas which were then injected into the crust above or poured out onto the sea bed as lavas.

Once Hess' idea was combined with Vine and Matthews' suggestion of the source of the anomalies, a hypothesis was born which was capable of being tested. Rather than a haphazard arrangement of positive and negative anomalies, it was now realised that they should be <u>symmetrical</u> about the ocean ridges. As each successive injection of magma occurred, it would displace earlier rocks symmetrically to each side. From time to time the earth's field would change polarity and some of the rocks would be reversally magnetised, resulting in alternate positive and negative anomalies, again centred symmetrically about the axis of the ridge. Figure 3.12 shows in diagrammatic form how this might happen. Detailed surveys of portions of oceanic ridges were then

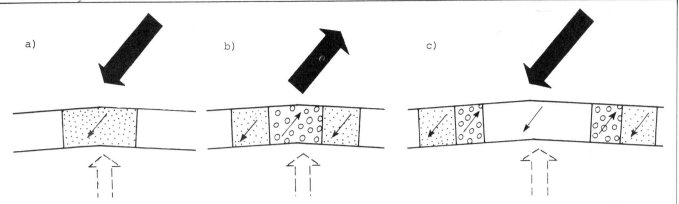

Fig. 3.12 a) First intrusion at ridge axis at time of normal polarity. Remanent magnetisation acquired in same direction as present field. b) Second intrusion at ridge axis at time of reversed polarity. First intrusion pushed apart. New material reversally magnetised. c) Modern intrusion at ridge axis under normal polarity. Earlier intrusions pushed apart. New material normally magnetised.

carried out and the predicted symmetry in the magnetic anomalies soon became apparent. A particularly clear example is found over the Reykjanes ridge, south of Iceland (Fig. 3.13). When the anomalies are plotted using the now conventional black for positive ones and white for negative, they give to the magnetic maps of the ocean floors the appearance of a zebra crossing and the magnetic anomaly belts have become known as 'sea-floor stripes'. The

hypothesis that the oceanic crust is generated at the ocean ridges and pushed along like a conveyor belt is known as the 'sea-floor spreading hypothesis'.

Soon after the initial ideas were formulated, it became possible to time the rate of sea-floor spreading. It was noticed that the width of each anomaly was proportional to the length of time determined for each

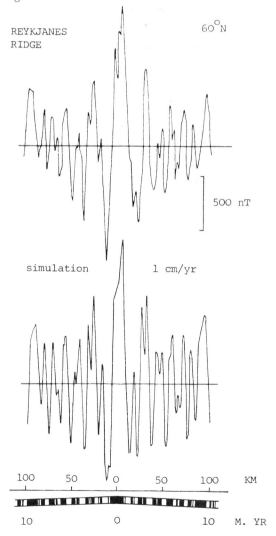

Fig. 3.13 Magnetic anomalies over the Reykjanes Ridge south west of Iceland, showing 'magnetic stripes' (positive anomalies black) and their bilateral symmetry.

Fig. 3.14 An observed aeromagnetic profile across the Reykjanes Ridge, compared with a computed profile assuming the reversal time scale shown below the profile.

133

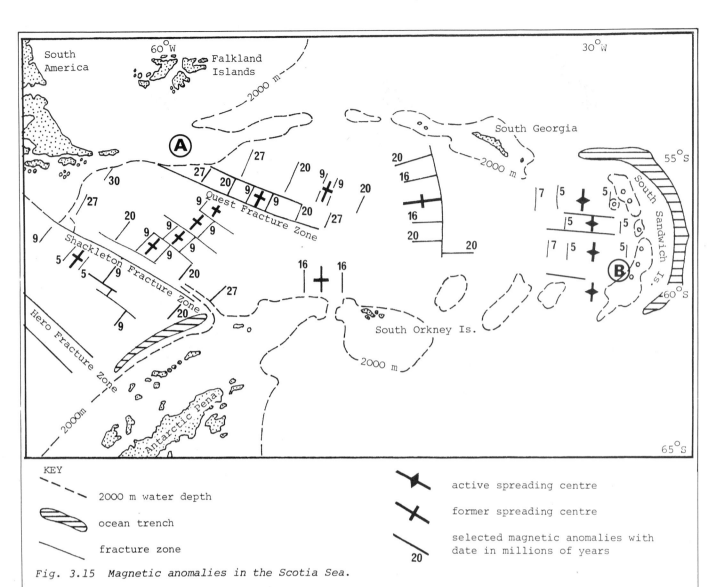

KEY

- - - -   2000 m water depth

ocean trench

fracture zone

active spreading centre

former spreading centre

selected magnetic anomalies with
date in millions of years

Fig. 3.15 Magnetic anomalies in the Scotia Sea.

period of normal or reversed polarity in
Iceland and elsewhere (Fig. 3.9).

Using this time scale, theoretical anomalies
were calculated and compared with the
anomalies actually measured across the
ridges (Fig. 3.14). In most cases the match
between calculated and measured anomalies
was very good and seemed to point to
spreading rates which have remained remarkably constant over long periods of time for
each of the major oceanic ridges. Thus,
in the North Atlantic, the spreading rate
has been a consistent 2 cm per year (i.e.
1 cm per year for each side of the ridge)
for a period of several millions of years.
In the East Pacific Rise, the rate is
higher, being about 10 cm per year.

Such was the rapid acceptance of the sea-floor
spreading hypothesis that it was followed by
a flurry of activity among sea-going
geologists. Existing records which had only
been partly interpreted were re-examined and
fresh surveys were conducted in critical
areas of the oceans. It is now appreciated
that each of the 'active' oceanic ridges (i.e.
ridges where there is also volcanic and
earthquake activity) exhibit the characteristic symmetrical magnetic anomalies and the
hypothesis is firmly established in geological thinking.

The Scotia Sea

The Scotia Sea provides just one example of
the way in which the sea floor spreading
hypothesis has affected our interpretation of
geophysical data.

The Scotia Sea lies within the Scotia Arc - a
chain of islands and submarine ridges which
appear to continue a structural link between
South America and Antarctica (Fig. 3.15).
Prior to 1963, four summer seasons had been
devoted to a magnetometer survey of the Sea.
The only analysis which could usefully be
attempted was statistically to define
magnetically 'quiet' areas, in contrast to
more 'active' ones. Thus Area A (Fig 3.15)
showed few anomalies and was equated with a
great thickness of sedimentary rocks forming
the sea bed. By contrast, the sea floor near
the South Sandwich Islands (Area B, Fig. 3.15)
contains many sharp anomalies, and was
regarded as volcanic in origin. Little could
be done to relate together the various
aspects of the Arc, nor to explain its
geological history.

After the publication of the sea floor
spreading hypothesis, subsequent expeditions
to the Scotia Sea have looked for, and found,
symmetrical anomalies similar to those about
the active ridges of the major oceans. A

typical magnetic profile is shown in Fig. 3.16a. The main map (Fig. 3.15) shows the positions of some more selected anomalies, with the dates in millions of years, derived from the reversal time scale in Fig. 3.9. The most striking feature is an active spreading centre in the east Scotia Sea, but there are other older and now inactive spreading centres in the central and western parts of the area. In addition, several major fracture zones have been discovered from the magnetic anomalies.

The full interpretation of the data is still not complete, but much has already been learned about the geological history of the region and the timing of the break-up of Antarctica, South America and Australia from the super-continent of Gondwanaland of which they had once been part.

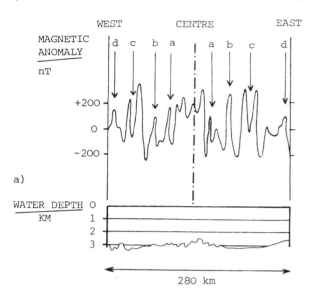

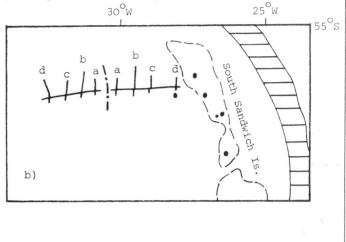

Fig. 3.16 a) Magnetic and bathymetric profile in the eastern Scotia Sea. Anomalies which are symmetrical about the central ridge are lettered. b) Location of profile and major anomalies shown in a).

4
The Seismic Method

For a variety of economic purposes, aero-magnetic and gravity surveys are used early in the exploration of a region, usually because of the speed of the survey and the relatively low cost. Unlikely areas may thus be eliminated and the search concentrated on potentially fruitful underground structures. It is then that the more expensive seismic techniques come into use. Here, explosives or other man-made sources release energy into the ground and sensitive detectors record the return to the surface of reflected or refracted waves.

Much of our knowledge of the interior structure of the earth comes also from studies of naturally occurring earthquakes, but first we shall examine some of the methods of seismic exploration, using controlled sources of energy.

Seismic Surveying

Reflection techniques

An everyday example of a sort of 'seismic' reflection technique is the ship's echo sounder, where a shock wave is transmitted several times a second from a transducer mounted in the ship's hull. The resulting shock wave travels out in all directions through the water, but part of the energy is reflected back to the ship from the sea bed and the time taken is automatically recorded. The velocity at which the wave travels through the water is known from many previous measurements and so the distance travelled by the wave may be calculated. In practice, this is done electronically, and the instrument usually produces a visual trace of these variations in water depth.

The principle of seismic reflection surveying is precisely similar, except that the 'reflector' for which we are looking is not the junction between sea water and the loose sediment on the sea bed, but rather is the junction between two or more sets of beds. A reflection survey will normally be designed to record many different reflectors, usually major bedding planes lying at considerable depth. Therefore more penetration is required and so more powerful energy sources are needed than for the echo sounder. The use of a lower frequency source enhances the depth of penetration of the waves of energy.

Energy sources

At one time, explosives were virtually the only source of seismic energy, but there is now a considerable variety of energy sources available. Most of these are safer and cheaper and allow a more rapid 'rate of fire'. The main methods in use are summarised in Table 4.1.

TABLE 4.1 - ENERGY SOURCES IN SEISMIC REFLECTION SURVEY

ENERGY SOURCE	NATURE OF SOURCE	AREAS OF USE	APPLICATION
AIR GUN	Sudden release of highly compressed air	at sea	Structure of deep sedimentary layers for possible oil and gas traps
EXPLOSIVES	Explosion produced by specially developed geophysical explosives	land and sea	As above, but used less now, for safety reasons
'VIBROSEIS' (Trade mark of Continental Oil Company)	Continuous vibration produced by electro-mechanical means	on land	As above, used particularly in built-up areas to minimise disturbance.
SPARKER	Electric arc discharge, like a giant spark plug	at sea	Sediment on sea bed and at shallow depth below
BOOMER	Metal plate rapidly repulsed from electro-magnetic coil	at sea	As above

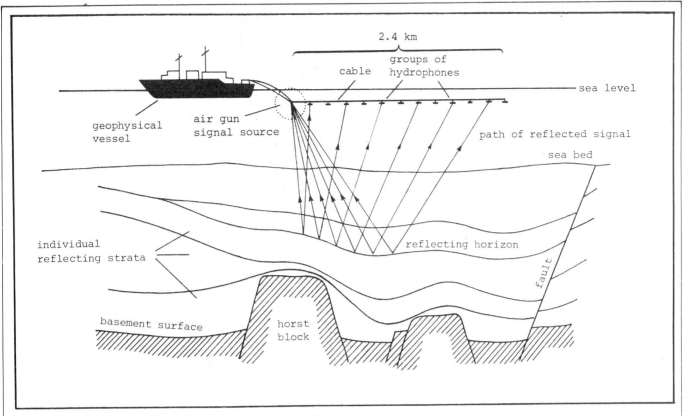

Fig. 4.1 *Diagram of offshore seismic reflection method.*

Detection equipment

Some of the energy which is reflected back
to the surface is picked up by small
detectors known as geophones (hydrophones at
sea). These either employ a moving-coil or
moving-magnet system or else a piezo-electric
crystal. Before the advent of computers, 20
or so geophones would be embedded in the
ground for each shot and the reflected
'arrivals' of energy picked off a paper trace
'by hand'. Most of the world's long-
established oil-fields were first discovered
by this means but since the early 1960s
computers have been programed to facilitate
the rapid analysis of data from many more
geophones. For example, at sea a survey
vessel tows both an air gun source and a
long 'streamer' of hydrophones, inserted into
a buoyant cable. The streamer may be as long
as 2.4 km, containing 48 recording sections
of cable, each one packed with a group of
hydrophones connected in parallel (Fig. 4.1).

The vessel can steam continuously on a set
course, 'firing' the air gun every few
seconds and recording the arrivals of
reflected energy on magnetic tape on board
the ship itself. Many reflections are thus
obtained from the same points on each bed
beneath, and these are 'stacked' by the
computer. This effectively reduces errors
from false reflections, such as a wave which
has bounced back and forth several times
between the reflecting bed and the surface.

The final result is compiled by the computer
and presented as a visual print-out.

North Sea oil and gas fields

The cover of this Unit shows a seismic
reflection profile some 18 km long in the

North Sea. Subsequent work proved the
existence of a hydrocarbon field known as
the Beryl Oil Field which is commercially
viable. The seismic record clearly shows the
anticlinal shape of the reflecting horizons,
representing different strata below ground.

It should be noted that the visual record is
presented as a sophisticated 'graph' of
horizontal distance versus reflection time.
It does not give the actual depth of each
reflecting layer. This can only be done
when we know the velocity at which the energy
has travelled through each layer. In the
first instance, these velocities are derived
from advanced computer analysis of the data,
or from specially placed extra shots.
Subsequently, these may be confirmed by borehole logs, if the area is drilled.

Most of the natural gas and oil reservoirs of
the North Sea have been located, since 1964,
by the use of such methods. No geophysical
techniques can actually prove the presence
of gas or oil: only a borehole can do that.
Even today, only one or two boreholes in ten
actually find any oil or gas, but, nonetheless, the 'dry' holes make an invaluable
contribution to the understanding of the
structure and stratigraphy of a region. As an
example, we can follow the fortunes of the
holding company of the Beryl Field, Mobil
North Sea Ltd.

An initial aeromagnetic survey was carried
out over the North Sea, including Block 9/13,
selected by the company. Fig. 4.2 shows the
location of the block. The aeromagnetic map
extract is printed as Fig. 3.6, with the
limits of Block 9/13 outlined. There is a
positive anomaly within the block, indicating
that magnetic 'basement' rocks are nearer the
surface here than in the immediate

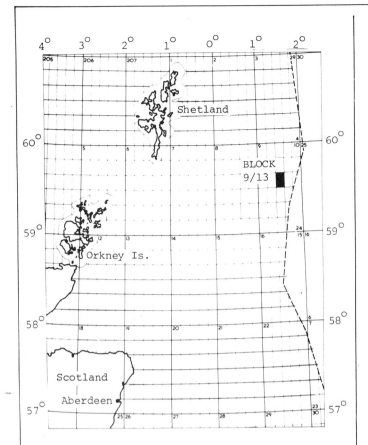

Fig. 4.2 Location of Block 9/13 in the North
Sea. The dotted line is the limit of the
British sector.

surroundings, possibly having been brought
up by block faulting. Calculations of the
approximate depth to the 'basement' may be
made.

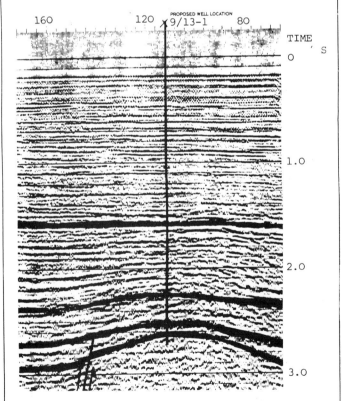

Fig. 4.3 Detailed seismic profile of the Beryl
Prospect in Block 9/13. Solid lines have been
drawn to emphasise major reflecting horizons and
a fault. Numbers at the top are shot points.
Times are for the double journey of the shock
waves (outward and reflected).

The magnetic survey was followed by seismic
reflection profiling, the results of which
are depicted on the cover. Fig. 4.3 is a more
detailed profile of the Beryl Field with
possible geological boundaries sketched in
and a fault inferred on the western side of
the structure. This is as far as the geophysicist can go: the next step is for the
geologist to provide the most likely geological interpretation regarding the ages and
types of rock present. At first this was
achieved by extension of the known geology
of the adjacent land areas, but was rapidly
improved once borehole programmes were
begun in other parts of the North Sea. Figure
4.4 is a sketch of the expected geology of
the Beryl structure, with the location of
the first 'wild-cat' well.

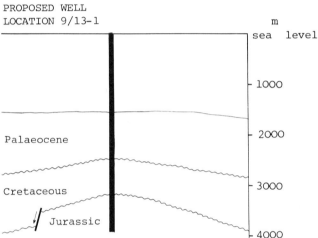

Fig. 4.4 Geological interpretation of the Beryl
Prospect in Block 9/13. Length of section
about 8 km. Intended total depth of
exploration well 3810 m.

In this case, the painstaking work was
rewarded. Almost 5 years after the project
was planned the well struck oil at 3002 m,
with more beneath. The well was stopped at
3148 m. Initial tests were favourable and a
production platform was set up to exploit
other parts of the field by further boreholes.

Even at this stage, some surprises may occur.
Figure 4.5 shows the results of drilling a
deflected well from a platform on the Dunlin
Field, owned by the Esso Petroleum Company
Ltd. The well logs show that the structure
revealed by seismic work is not quite such a
favourable reservoir for oil as had been
thought.

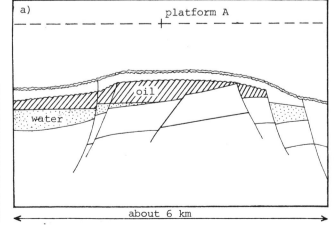

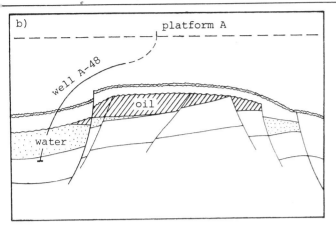

b) platform A

well A-48

oil

water

Fig. 4.5 The sectional diagrams show a) the way in which the oil and water deposits were assumed to be lying in the Dunlin reservoir and b) how the interpretation changed after well A-48 was drilled. This well revealed that the westernmost fault block was lower than had originally been thought.

Refraction techniques

At first sight, the principle of surveying by seismic refraction methods does not seem so obvious as the reflection technique, but in practice, refraction surveys are of great value. Shallow refraction work, using small explosive charges, a sledgehammer, or a 'thumper' (a heavy weight dropped from a lorry) is used in site surveys for foundation work in major civil engineering projects. Deeper crustal studies may also be made, and under favourable conditions, the entire thickness of the earth's crust may be determined by using charges of up to several hundred kg of explosives.

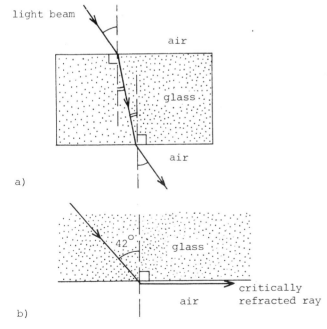

light beam

air

glass

air

a)

42°   glass

air   critically refracted ray

b)

Fig. 4.6 Refraction of light through a glass block.

The principle of the refraction of waves of energy will be familiar to anyone who has experimented in school physics with a beam of light and a glass block. The light beam can be seen to bend as it enters and leaves the block (Fig. 4.6a). If the angle of incidence of the light beam is altered, a point is reached where it fails to emerge from the other side of the block; it has been refracted at the critical angle and then travels parallel to the side of the block. For a glass block in air, the critical angle is about 42°, depending on the type of glass (Fig. 4.6b).

A very similar thing happens in the earth when various layers occur, one on another. Different rock types transmit shock waves at different velocities, just as light travels more slowly in glass than it does in air. In nature, velocities at which shock waves are transmitted through rocks generally <u>increase</u> as the wave travels deeper. Where deeper strata are of lower 'seismic velocity' than those lying above, they are very difficult to detect.

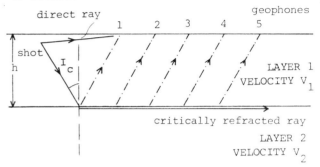

direct ray   geophones

1 2 3 4 5

shot

h I_c LAYER 1 VELOCITY V_1

critically refracted ray

LAYER 2 VELOCITY V_2

Fig. 4.7 Refraction of shock waves in the surface layers of the earth.

The basic principle is shown in Fig. 4.7. Waves from the explosion travel out in all directions, like the ripples on a pond when a stone is thrown in. The passage of the wave front in any direction is shown diagrammatically by an arrow. Some energy travels near to the ground surface (the direct wave), some is refracted into the next layer and the energy is eventually dissipated. One wave path, however, meets layer 2 at the critical angle (I_c) and is refracted along the top of layer 2, travelling at the new velocity V_2.

Energy is continually being returned into Layer 1 and this can be detected by the geophones. The time between the shot and the first arrival of the shock waves at the geophones may be measured very accurately (usually to better than a millisecond). At first, the direct wave will be the first to arrive, but as the geophones are spread further from the shot, the critically refracted wave will be received. Usually, V_2 is greater than V_1 and so eventually the refracted wave is the first to arrive, since it has covered most of its journey at the faster speed. This principle is continued with arrivals from deeper layers. Arrival times are plotted against the distance between the shot and the geophones (Fig. 4.8).

Such a graph is known as a 'travel-time plot' and the velocity of seismic wave transmission in each layer is easily obtained from the graph. In the form in which it is plotted in Fig. 4.8, V_1 is obtained from the reciprocal of the gradient of the line XY and V_2 similarly from the line YZ. Try measuring V_1 and V_2 for yourself.

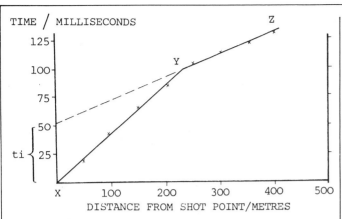

Fig. 4.8 'Travel-time' plot for a two-layer seismic refraction profile.

'Outcrop shooting' on rocks of all types has enabled us to determine characteristic (but not unique) values for seismic wave velocities. Some typical figures are:

Dry sand/gravel	500-1000 m s^{-1}
Shale	1000-3000 "
Sandstone	2000-4300 "
Rock Salt	4700-5700 "
Slate	4700-5300 "
Basic igneous rocks	up to 7700 "

Which of these rocks types could be present for the survey shown in Fig. 4.8?

In addition to the velocities in the layers, it is also possible to calculate the layer thicknesses. The appropriate equation for Layer 1 (derived ultimately from Snell's Laws of Refraction) is:

$$\text{Thickness } h = \frac{ti}{2} \frac{V_2\, V_1}{\sqrt{V_2{}^2 - V_1{}^2}}$$

where ti is the 'intercept time' in seconds (i.e. 0.052 seconds in Fig. 4.8)

Calculate the thickness of Layer 1. You cannot of course deal with Layer 2 until the travel-time curve indicates another change in velocity as the next layer down is reached.

Two examples of the application of the seismic refraction method illustrate the versatility and some of the uncertainties of the technique.

Hammer seismic profile

A seismic line was 'shot' across the inside of a meander bend by a party of sixth formers, the equipment used being a simple seismic set with one geophone. The shock-wave was produced by the most muscular student striking an iron plate with a sledge hammer. The moment of impact was recorded from an impact-switch attached to the hammer, thus enabling travel times of the wave to the geophone to be electronically recorded. A sketch map of the locality and the travel-time plot is shown in Fig. 4.9.

Compute V_1 and V_2 and the thickness of Layer 1. Then attempt your own brief geological

interpretation, using, in addition, the data shown on the sketch map. Assume the refractor is horizontal. A suggested answer is given at the end of the Unit.

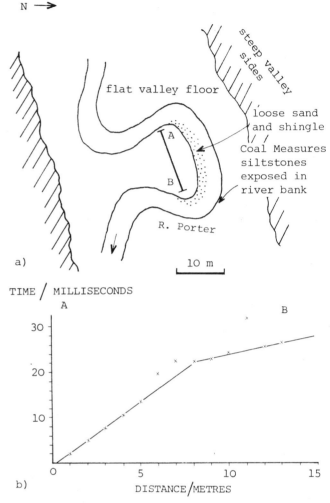

Fig. 4.9 a) Sketch map of a section of the River Porter, Sheffield (SK 312852) showing location of seismic refraction profile AB. b) Travel-time plot for the line AB.

Seismic refraction survey of Cardigan Bay, Wales

During the 1960s, the University of Birmingham carried out a research project involving seismic refraction and 'sparker' surveys from ships in Cardigan Bay. The survey was initiated to try to solve an interesting problem.

Earlier gravity surveys of West Wales had shown a rapidly decreasing Bouguer anomaly towards the coast (see Fig. 4.10). This was assumed to be due to low density rocks lying along the coast and beneath the Bay and the gravity gradient was so steep as to suggest a faulted boundary parallel to the coast. What could be the cause of the gravity 'low'? Reference to a copy of the 10 mile map of the country will show that much of the geology of West Wales comprises Lower Palaeozoic rocks, mostly greywackes, shales and slates. Large granitic bodies are absent: so too are major deposits of Mesozoic or younger rocks, either of which could reasonably be expected to provide the necessary density contrast. Most of the palaeogeographic maps available

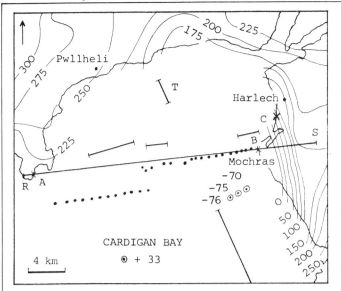

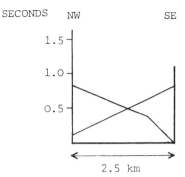

KEY

— 10 Bouguer anomaly in g.u.

⊙ gravity station on reef
Bouguer anomaly in g.u.

├──┤ sea seismic line

· shots to land seismic
stations A, B, C.

Fig. 4.10 Gravity anomalies and seismic stations in part of Cardigan Bay.

Fig. 4.11 Travel-time graph for line T.

of the seismic results will allow; the geological meaning of the profiles is open to discussion and, as is usual with such work, the research workers presented their findings to a learned society in two papers.

Table 4.2 summarises the ideas put forward by the authors and some of the suggestions made by other speakers in the discussion which followed the lectures.

Clearly, there was considerable diversity of opinion, much of it coming from closely reasoned geological arguments from geologists well versed in the rocks of Wales. The Birmingham group's second paper concluded with the comment that a borehole at Mochras, on the westernmost part of the coast would clinch the argument.

A few years later, just such a borehole was undertaken. The results (Fig. 4.13) were quite a surprise. It now seems that much of 'Layer 1' is of middle Tertiary age, a time which is sparsely represented in the rocks of mainland Britain! Furthermore, Layer 2 corresponds to the most complete succession of Liassic (Lower Jurassic) rocks anywhere in the country!

The implications of the combined geophysical and borehole studies are far reaching. First, sedimentary basins of Mesozoic and Tertiary age are usually the most likely reservoir rocks for oil and gas. Commercial exploration for oil and gas is now being carried out in the seas west of Britain.

in the 1960s interpreted West Wales and Cardigan Bay as having been land areas from the end of the Silurian to the Cretaceous (e.g. Fig. 4.14a).

On the basis of this brief survey and the 10 mile geological map, can you suggest possible causes for the trend of the gravity anomalies, before you read further?

The seismic refraction survey was carried out in the northern parts of the Bay and was later followed by sparker surveys to delimit the near-surface sediment. Where possible, each refraction line was 'reversed' to eliminate errors due to the dip of the strata. A typical travel-time graph is shown in Fig. 4.11.

The results of several seasons' work are summarised in the interpretation in Fig. 4.12. This is as far as direct interpretation

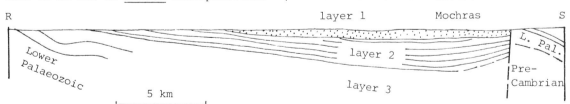

Fig. 4.12 Interpretation of the structure of part of Cardigan Bay from geophysical data.

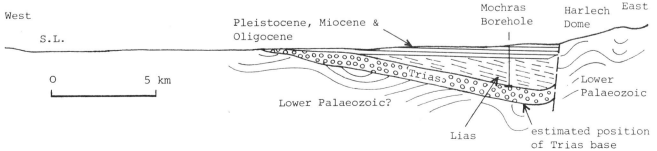

Fig. 4.13 Section through Tremadoc Basin and Mochras boring.

TABLE 4.2 – GEOLOGICAL INTERPRETATIONS OF GEOPHYSICAL WORK IN CARDIGAN BAY

PROBLEM	SOLUTIONS SUGGESTED BY BIRMINGHAM UNIVERSITY GROUP	DISCUSSION AT TWO MEETINGS OF THE GEOLOGICAL SOCIETY OF LONDON
NATURE AND AGE OF LAYER 1	Post-glacial material overlying Tertiary and/or Mesozoic clays. (Jurassic fossils occur in beach pebbles. Tertiary clays crop out in Northern Ireland)	a) Whole of Layer 1 could be glacial material, by analogy with Scottish example (K Dyer) b) Could be Trias, similar to Midlands (T N George). c) New aeromagnetic survey shows few anomalies in Cardigan Bay. Could be Trias or Tertiary (W Bullerwell).
NATURE AND AGE OF LAYER 2	a) Ordovician sediments. (Known on mainland. Seismic velocities consistent with outcrop shooting on Ordovician) b) Could be Permo-Trias or Carboniferous, but unlikely.	a) Carboniferous, on palaeo-geographic grounds. (T N George) b) Trias - similar to other Triassic basins in Cheshire (O T Jones). c) If junction between Layers 2 and 3 is really the base of the Ordovician, why is it not folded, as on the mainland (W F Whittard)?
NATURE AND AGE OF LAYER 3	Cambrian (known on mainland).	No comment.
NATURE OF EASTERN BOUNDARY OF CARDIGAN BAY	Major fault (from steep gravity gradient and seismic work on land).	Not faulted. Prefer an overlap junction (T N George).
GENERAL STRUCTURAL SETTING OF CARDIGAN BAY	a) Structural basins preferred. b) Buried granite beneath surface sediment. Top of granite would be at 7 km. depth and base at 26 km. Possible from gravity anomaly but unlikely.	a) New sea-borne gravity profile shows 600 g.u. negative anomaly. Consistent with deep sedimentary basin or buried granite (M H P Bott). b) Could be late Mesozoic/Tertiary downwarp as known in western English Channel (S E Hollingworth).

Secondly, our ideas of the palaeogeography of the country at various times in the past must be completely redrawn. Figure 4.14a shows the previously proposed palaeogeography for the Lower Jurassic and Fig. 4.14b a more recently drawn one in the light of offshore work such as that described above. Quite a lot of land seems to have sunk since 1951!

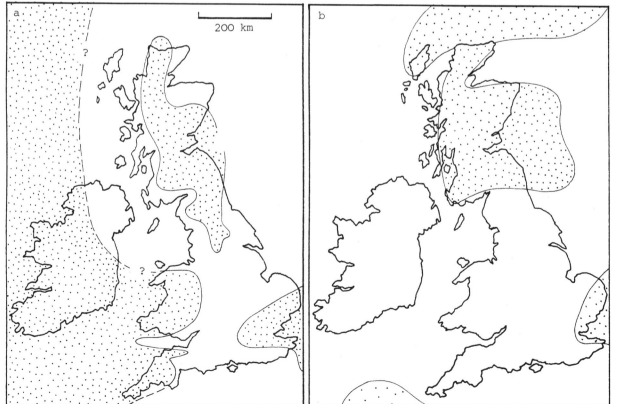

Fig. 4.14 Two interpretations of the palaeogeography of the British Isles during Lower Jurassic times a) after L J Wills 1951 b) after A Hallam and B W Sellwood 1976. Land areas stippled.

Earthquake Seismology

General principles and wave types

Broadly speaking, the principles involved in employing earthquake energy as a seismic source are similar to those of explosion seismology. Shock waves are sent out in all directions following an earthquake and they are refracted as they pass through layers with different seismic velocity characteristics. Because the natural energy produced is so much greater than that from man-made sources, waves from a strong earthquake actually travel right through the earth and can therefore provide information about the deep interior. The time of arrival of the shock wave as it emerges after its journey is recorded on a seismograph, which acts in the same manner as an elaborate geophone. The seismograph converts the shock wave arrivals to electrical impulses which, in turn, are recorded as 'wiggly line' traces on a moving-drum recorder. Accurate time marks are also made on the recorder, so that the precise time of arrival of each seismic 'event' can be read later.

To make full use of earthquake seismology, we need to study more closely the nature of the shock waves. For most straightforward purposes in seismic surveying, using man-made sources, we can regard the waves as being of one type, but this is not true of earthquake seismology. Earthquakes produce several different types of wave, which travel in different ways and with different velocities.

Body waves (P and S waves)

Those which travel through the earth are known as 'body-waves' and comprise two types. The faster-travelling ones are known as P-waves (or longitudinal waves) and they are propagated by a series of compressions and rarefactions of the particles in the earth. In a way, they may be likened to a string of railway trucks being shunted. As the engine hits the first truck, the movement is passed on to the second truck, but the first one rebounds. This continues down the line until the furthest truck moves. After the train gets under way the analogy is no longer valid, since all the wagons are moving together and not just oscillating backwards and forwards over one spot.

The slower body waves are known as S waves (or transverse waves). These are transmitted through a medium by the motion of particles at right angles to the direction in which the wave is travelling.

Surface waves (L Waves)

Two other types of wave are also produced. These do not travel deep in the earth, but are restricted to the surface layers. Such surface waves are known collectively as L waves and comprise Love waves which are transverse waves and Rayleigh waves, which travel with an elliptical particle motion. L waves are capable of being transmitted right round the globe and it is these waves which are responsible for the ground-roll which produces such devastation during an earthquake. The P, S and L waves appear on the seismograph record as shown in Fig. 4.15.

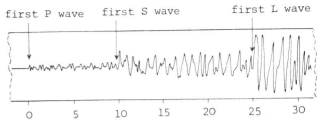

Fig. 4.15 *Part of seismogram for an earthquake some 8500 km from the recording station.*

The knowledge of wave types is of immense practical interest for several reasons. First, they can be used to enable the distance between the earthquake and the observatory to be calculated. There is fortunately an empirical relationship between the time of arrival of the P, S and L waves and the distance from the source of energy. This has been plotted from 'known' earthquakes and is shown in Fig. 4.16.

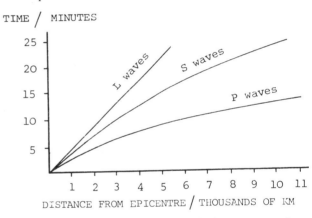

Fig. 4.16 *Time-distance graph for P, S and L waves.*

Calculations from the records of the same earthquake made at three or more seismograph stations enable the position of it to be pinpointed. The position may be plotted on a globe as the epicentre (that is, the point on the earth's surface directly above the actual centre or focus of the event itself).

Secondly, passage of the P and S waves through the earth depends upon the elastic properties of the medium through which the waves are passing. The relationship may be measured in the laboratory under controlled conditions and then applied to the earth's interior. Thus, P wave transmission is related to the properties of the medium as follows:

$$P \text{ wave velocity} = \sqrt{\frac{K + \frac{4}{3}\mu}{\rho}}$$

Where K is the bulk modulus, or incompressibility of the medium; μ is its rigidity modulus (related to its shear strength) and

143

ρ is its density. The corresponding equation for S waves is:

$$\text{S wave velocity} = \sqrt{\frac{\mu}{\rho}}$$

Given that K is always positive, and that μ is zero for a liquid, try to answer the following:
1. *Which is always greater, P wave or S wave velocity?*
2. *Which type of wave will <u>not</u> be transmitted by a liquid?*
3. *How does elastic wave velocity depend on density?*

Your answer to 3 may have surprised you, since you probably expected a <u>directly</u> proportional relationship, not an inverse one. In fact, densities do increase with depth and so do wave velocities, but this is because the elastic moduli increase faster and more than counteract the rise in density.

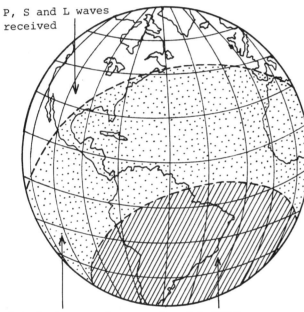

"shadow" zone where only L waves are received

P and L waves received but no S waves

Fig. 4.17a The 'shadow zone' from an earthquake in Japan.

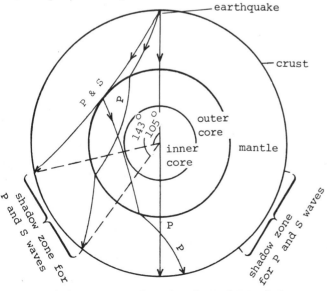

Fig. 4.17b The paths of selected P and S waves through the earth.

The deep structure of the earth

When the arrivals from many different earthquakes are plotted, some rather surprising results emerge. From any one earthquake, stations in some parts of the world record complete sets of P, S and L waves; others record P and L waves only, whilst others still receive only L waves. The distribution is not haphazard, but follows the pattern shown in Fig. 4.17a. The area where only L waves are recorded is known as 'shadow zone' and spans the earth, at an angular distance from the earthquake of between 105° and 143°. Beyond this zone, P waves are received again, but no S waves.

Our knowledge of the elastic moduli leads us to infer that at least part of the earth's interior must be liquid and that the S waves have been absorbed. The P waves have emerged after being progressively refracted on their passage through the earth. Figure 4.17b shows an interpretation of the wave paths through the earth and Fig. 4.18 is a plot of depth from the surface against wave velocity. There is a marked drop in the velocities at 2900 km depth and this is taken to mark the boundary between the mantle and the liquid core. The boundary is known as the Weichert-Gutenberg Discontinuity, after its discoverers.

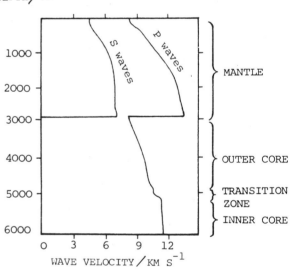

Fig. 4.18 Velocity/depth curves from the earth's surface to its centre (details of the crust omitted).

Deeper within the core, the P wave velocities begin to rise again. Theoretical considerations of the likely properties of materials at such great depths lead us to believe that this inner core is solid.

The nature of the crust from earthquake and explosion seismology

The thickness of the crust

Evidence for the nature of the uppermost layers of the earth mostly comes from near-earthquakes, which can be regarded in a similar way to man-made explosions.

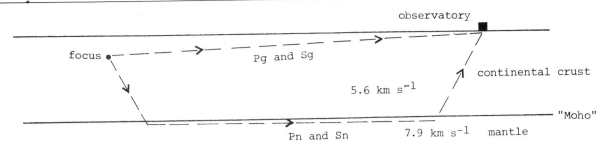

Fig. 4.19 Crustal structure determined from near earthquake.

A Yugoslavian named Mohorovičić was the first to notice <u>two</u> sets of seismic waves arriving at the seismographs from a single nearby earthquake shock which took place in 1909. He realised that the first waves to arrive represented energy which had been <u>refracted</u> at a depth of over 50 km below the surface and had travelled most of the journey in a higher velocity medium (P_n and S_n in Fig. 4.19). The second arrivals were produced by slower ground waves which had travelled direct to the observatories through the crustal rocks (P_g and S_g in Fig. 4.19). The refractor responsible seemed to represent a sharp junction separating the rocks of the crust from the underlying mantle and it soon became known as the 'Moho', or 'M-Discontinuity' after its discoverer. He calculated an average P wave velocity for the continental crust of Yugoslavia of 5.6 km s^{-1} and for the upper mantle beneath of 7.9 km s^{-1}.

Subsequent work in many parts of the world has proved the widespread occurrence of the M-Discontinuity, although it seems that in some areas it represents a gradual transition zone rather than an abrupt break. The seismic velocity of the uppermost mantle also varies, but, apart from some anomalous areas, its minimum value is 7.7 km s^{-1} for P waves, which is appreciably higher than that of most crustal rocks.

Near-earthquake and large-scale explosion seismology have also been applied to test the hypotheses of Airy and Pratt regarding the variations of thickness and density of the crust (see p. 8). It soon became clear that there are major contrasts between the oceanic crust and that of the continents. Such seismic work has shown that, as Airy predicted, the oceanic crust is both thinner and denser than the continental crust, the average thicknesses being 11 to 12 km (inclusive of sea water) and 35 km respectively.

Beneath many mountainous areas, the crust is usually much thicker than average. For example, for the Rocky Mountains it is over 60 km thick. This also seems to vindicate Airy's hypothesis, but more recent seismic work has shown that there are some regions, such as the Basin and Range Province of the western USA, where the crust is continental in type and yet only 25-30 km thick. Here, the observed gravity anomalies are more readily explained by lateral changes in rock density, as invoked by Pratt.

Structure of the oceanic crust

Not only the thickness, but also much of the detailed structure of the crust can be determined by seismic means.

The oceanic crust is rather less variable than that of the continents. It is also less troublesome to detonate large explosions at sea rather than on land. (Even so, 150 kg of TNT may be required to produce measurable refractions from the M-Discontinuity at sea.) Fig. 4.20 gives a typical result from refrac-

TABLE 4.3 - THE OCEANIC CRUSTAL LAYERS

LAYER	P WAVE VELOCITY (KM S^{-1})	AVERAGE THICKNESS (KM)
Sea Water	1.5	4.5
Layer 1	1.6 - 2.5	0.4
Layer 2	4.0 - 6.0	1.5
Layer 3	6.4 - 7.0	5.0
Upper Mantle	7.4 - 8.6	

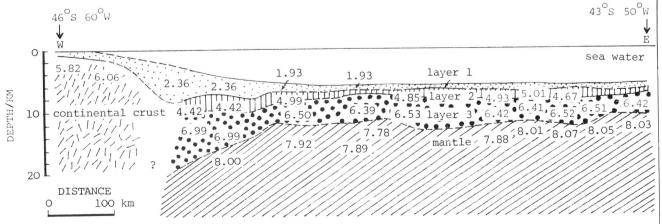

Fig. 4.20 Oceanic crustal structure determined by seismic refraction in the Atlantic ocean east of Argentina. Figures refer to seismic velocities in km s^{-1}.

tion surveys in the South Atlantic Ocean. The compilation in Table 4.3 is a section through an 'average' oceanic crust, derived from many such surveys in all the major oceans. Seismic reflection work is also used to provide additional detail of the nature of Layer 1.

Geologically, the following interpretation seems most likely:

Layer 1: unconsolidated sediments passing downwards into consolidated sediments.

Layer 2: basalts, or consolidated sediments, probably the former.

Layer 3: partly metamorphosed basic igneous rocks.

Structure of the continental crust

Although it is more difficult to conduct deep seismic experiments on the continental crust, many have been undertaken in recent years. An example is the Lithospheric Seismic Profile in Britain (LISB) of 1974, where refraction lines were shot from the North of Scotland to the English Channel and across into Northern France (Fig. 4.21). A typical velocity-depth interpretation of the crust below the Scottish Highlands is shown in Fig. 4.22. In this case, the survey team were lucky enough to record a minor earthquake which occurred below the coast of West Scotland whilst the equipment was running.

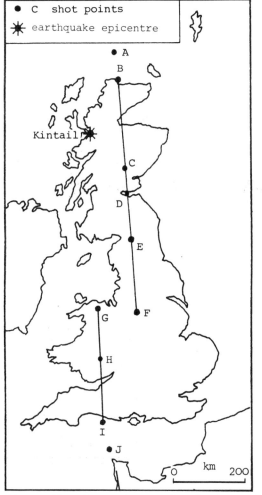

Fig. 4.21 Shot points and seismic profiles used during the LISB experiments of 1974.

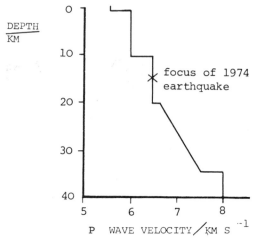

Fig. 4.22 Velocity/depth interpretation of data from the Kintail earthquake 1974.

A simplified version of the LISB team's interpretation of the depth of the M-Discontinuity beneath Britain is shown in Fig. 4.23. Much of it is close to the average depth beneath the continents of 35 km but there are significant variations. Compare this thickness with the average 11 to 12 km for the depth of the M-Discontinuity beneath the oceanic crust (Table 4.3).

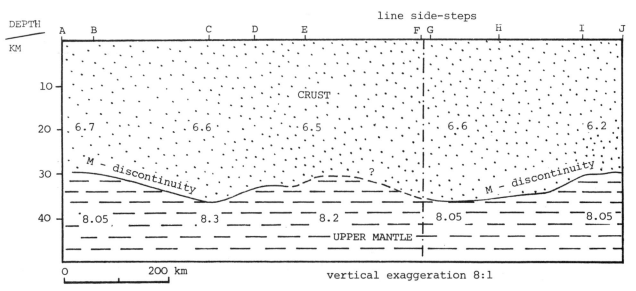

Fig. 4.23 Generalised crust-upper mantle section of the British Isles along the line A-J. P wave velocities are average figures for the whole crust and for the upper mantle and are in Km s^{-1}.

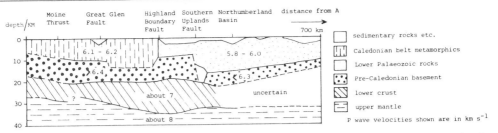

Fig. 4.24 Diagrammatic cross-section of the crust and upper mantle of northern Britain.

The LISB experiment has also resulted in a tentative 'model' for structures within the crust of Northern Britain, a simplified version of which is shown in Fig. 4.24. Several layers may be equated with the known sedimentary, igneous and metamorphic geology represented at outcrop. Beneath them lies the 'lower crust' which is of comparable velocity to Layer 3 beneath the oceans. Many workers believe that the geology of this 'lower crust' is broadly similar to that of Layer 3.

At one time, it was thought that a marked break in seismic velocities between an upper and a lower continental crust was always present. Indeed, in places where it could be measured the junction was referred to as the 'Conrad Discontinuity'. It is now clear, however, that there are very many regions where there is no sharp break within the continental crust and so the term is of doubtful validity.

The low velocity layer

Recently, careful examination of seismograph records has revealed the existence of a layer, or zone, within the upper mantle, where seismic velocities are actually lower than those of the overlying layers (Fig. 4.25). This is usually taken to mean that the layer must be physically weak or unusually plastic, and it is sometimes given the Greek name 'asthenosphere' to express this fact.

By contrast, the parts of the mantle lying above the asthenosphere together with the overlying crust, form a relatively brittle shell, known as the 'lithosphere', or 'rocky shell'. On average, the lithosphere is between about 100 km and 200 km thick. For many purposes, we can regard the crust and upper mantle which comprise the lithosphere as acting in unison, able to 'ride' as it were, over the weak asthenosphere.

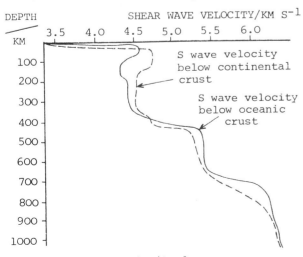

Fig. 4.25 The low velocity layer.

The surface distribution of earthquakes

So far we have only considered the information yielded by earthquake waves travelling through the earth, but there is also much to be learned from a study of the surface distribution of earthquakes. It is common knowledge that some parts of the world are more subject to earthquakes than others, although nowhere can be considered to be totally immune. The map (Fig. 4.26) shows the distribution of the main earthquakes and it is immediately clear just how restricted in occurrence they are.

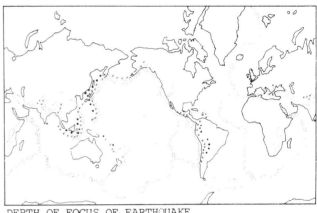

DEPTH OF FOCUS OF EARTHQUAKE
shallow (0-70 km)   intermediate (71-300 km)   deep (301-700 km)

Fig. 4.26 The distribution of earthquakes.

The most marked seismic belt runs three-quarters of the way around the margins of the Pacific Ocean. A similar line bisects the Mediterranean area and continues on through southern Asia and Indonesia to meet the circum-Pacific belt. Smaller belts occur in the Caribbean and Scotia Seas. Narrow belts of earthquakes run through the major oceans of the world, whilst the East African Rift Valley is marked by a further line of seismic activity. Other, minor earthquakes occur in association with isolated volcanoes such as the Hawaiian islands.

Noticeable though all these earthquake epicentres are, it must not be overlooked that by far the greatest part of the earth's surface is largely aseismic, that is, inactive. Why then are the earthquake zones so special? When we examine them more closely, we find that each major type of seismically active belt corresponds with a surprising number of other geological features. In list form these are as follows:

Circum-Pacific belt, Caribbean and Scotia Sea

1. Earthquake foci lie along a plane often inclined at about 45° to the horizontal.

In regions such as the Pacific coast of South America, shallow foci (0 to 30 km in depth) are on the seaward side. Intermediate ones (71 to 300 km in depth) lie roughly beneath the coastal areas, whilst deep focus events (301 to 700 km in depth) lie beneath more inland areas (Fig. 4.26).

2. Where the earthquake belt lies near to the junction between a continent and an ocean, the continental margin usually contains a young fold mountain range, frequently with active volcanoes.
3. Other parts of the belt, in the open ocean, are marked by island arcs, that is, chains of mostly volcanic islands arranged in a gentle curve, whose convex side is towards the middle of the ocean.
4. The dominant lava type in the island arcs and fold mountain ranges is andesite, a lava of intermediate composition.
5. A deep trench in the ocean floor usually lies off the convex side of the island arcs or off the coast of the fold mountain ranges.
6. Markedly negative gravity anomalies occur in association with these regions.
7. Heat flow outwards from the earth in such areas is abnormally high, i.e. there is a heat surplus being generated by some process deep in the earthquake zone.

Some of these observations also apply to the Mediterranean/Himalayan earthquake belt.

Mid-oceanic earthquake belts

1. Earthquakes are of shallow depth only (0 to 70 km depth).
2. The lines of earthquakes lie beneath major ocean ridges, which rise like underwater mountain chains some 2000 to 3000 metres above the general level of the sea floor.
3. The ridges are associated with vulcanicity of basaltic lava type.
4. Heat flow over the oceanic ridges is abnormally high.
5. The rocks of the sea bed exhibit magnetic 'sea floor stripes' as described in an earlier section.

Plate tectonics – a brief note.

We have already attempted to explain some of the features of the ocean ridges. Presumably the shallow earthquakes with which they are associated result from the generation of magma in the upper mantle and the 'pulling apart' of the lithosphere as the magma is injected or extruded. If this process alone were acting throughout the world, it would be necessary for the earth to be continually expanding to accommodate the new lithospheric material, but few geologists believe this to be the case. Instead, it now seems that we may look at the other major earthquake zones, such as the circum-Pacific belt, as the site of a kind of 'return mechanism' whereby at least some of the lithospheric rocks are returned to the mantle. The angle from the horizontal at which the earthquake foci are inclined, and the preponderance of features of compressional origin, suggest that one portion of lithosphere is being pushed down against another, e.g. where a continent meets an ocean, it is the material on the oceanic side which is pushed down (Fig. 4.27).

Whilst we have concentrated on all the exciting activity of the relatively limited earthquake belts, it must be remembered that most of the earth's surface is aseismic. This has led to the suggestion that the lithosphere is divided into a number of portions or plates, each of which may be moved about over the asthenosphere as a rigid unit, but is only subject to deformation at its junction with another plate. Most geologists now accept that the earth's surface is composed of some seven major lithospheric plates plus a number of smaller ones. Since the boundaries are all marked by earthquake zones, you might like to try dividing the world up into plates on a tracing of Fig. 4.26 and then comparing your answer with a map in a modern textbook such as *Volcanoes*, published by the I.G.S.

The implications of the plate tectonic theory, as it is known, are quite profound and are beyond the scope of the present book. You will, however, find that plate tectonic ideas now underpin much of your other geological reading, in many cases with good justification. In some instances, however, the connections are rather hazy, so you should retain your critical faculties as you read!

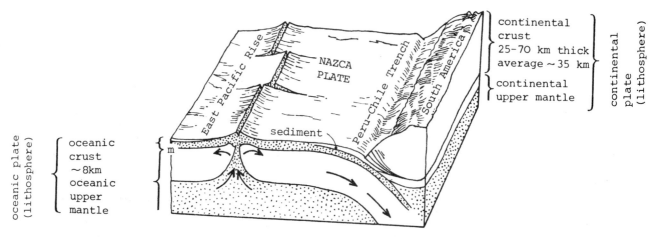

Fig. 4.27 Block diagram of the 'Nazca Plate' in the eastern Pacific.

Answers

Palaeolatitude of Great Britain (Table 3.1 p.14)

Present	$52°N$
Tertiary	$42°N$
Cretaceous	$38°N$
Jurassic	$31°N$
Triassic	$31°N$
Permian	$12°N$
Carboniferous	$0°$

It appears as though the region stood still between the Triassic and Jurassic, but it is quite probable that there was a change in longitude which cannot be measured from palaeomagnetic specimens.

Seismic refraction travel-time (p. 24)

V 2350 m s^{-1}: Could be shale or sandstone
V_2 5400 m s^{-1}: Could be slate or basic igneous rocks or rock salt
h 67.9 m

Hammer seismic profile (p. 24)

Your analysis of the travel-time curve should have shown:

$V_1 = 364 \text{ m s}^{-1}$, $V_2 = 1160 \text{ m s}^{-1}$, $h = 3.07 \text{ m}$

This could be interpreted in two main ways:

a) 3.07 m of river alluvium overlying Coal Measures siltstones.
b) 3.07 m of dry alluvium above the water table, with saturated alluvium below. Coal Measures siltstones not reached because the line was not long enough.

Note that the velocity of 364 m s^{-1} is close to that of the transmission of sound in air, but it is also known in loosely consolidated gravels.

Seismic wave velocities (p. 28)

1) P wave velocity
2) S waves
3) Velocity is _inversely_ proportional to density.

Further Reading

Dunning, F.W.
Geophysical Exploration, HMSO, 1970. Simple, well illustrated booklet produced by the Geological Museum.

Griffiths, D.H. and King. R.F.
Applied Geophysics for Geologists and Engineers. Pergamon Press, 1981 (2nd edition). Covers all the main methods of geophysical exploration.

Hallam, A.
A Revolution in the Earth Sciences. Oxford University Press, 1973. Historical survey of the growth of current ideas in global geology.

Holmes, A.
Principles of Physical Geology. Revised by D.L. Holmes, Nelson, 1978 (3rd edition).

Khan, M.A.
Global Geology. Wykeham Science Series. Taylor and Francis Group, 1976. General survey of the use of geophysics in global geology.

Igneous Petrology

Contents

Acknowledgements

We are grateful for permission to reproduce
the following illustrations:
Fig. 1.1, U.S. Geological Survey; Fig. 1.6a,
R.T. Holcomb, U.S. Geological Survey; Fig. 1.
6b, J.D. Griggs, U.S. Geological Survey; Fig.
2.1 Professor P.E. Baker; Fig. 2.2, The
Institute of Geological Sciences; Fig. 3.3
The Joint Matriculation Board; Fig. 3.11, Dr
J.E. Treagus.

The following figures have been based on
illustrations from the sources indicated:
Figs 1.2, 1.3, 1.5, Thomas, A.J. *Geology
Teaching Vol 6 No 2*; Figs 1.4, 3.4, 3.5, 3.6,
Internal Processes, Open University; Figs
3.19, 3.24, 5.3, Read, H. and Watson, J.
Introduction to Geology, Vol. 1.

Foreword

This book covers the general nature and
distribution of igneous activity, the classification of igneous rocks and the evidence
which the rocks themselves provide about
their origins. The main emphasis of the book
is on intrusive igneous activity; further
information on surface vulcanicity should be
sought in other texts, many of which contain
abundant illustrations of modern volcanoes.

Except where otherwise acknowledged, the
photographs in this Unit have been taken by
the authors. The scale bar on such photographs is 5 cm in length.

There are various exercises built into the
text. You should try to complete these,
indicated by italic type, before going any
further.

Units of measurement. The derived S.I. unit
of pressure is the pascal (Pa):
1 Pa = 1 newton per square metre
1 Kbar = 10^8Pa

1
Modern Volcanic Eruptions

Fig. 1.1 The eruption of Mount St Helens, May 18 1980.

Introduction

In March 1961 a little-known event took place in the remote South Atlantic. A ship had unexpectedly become stuck in the open ocean! What was the cause of this potentially dangerous episode? An underwater volcano had erupted on the sea-bed beneath and pumice rose to float on the surface. Pumice completely surrounded the ship and rendered her quite helpless by clogging the inlets to the water cooling system for the ship's engines. It was quite a while before the efforts of seamen with long poles combined with the freshening wind to free the vessel and enable her to resume normal duties.

Thousands of miles away, in 1980, another natural event occurred which was to horrify the world. Mount St Helens erupted. This volcanic peak in the western USA had begun to give notice of its intention when minor earth tremors started in March 1980. Minor eruptions followed and the authorities took carefully planned measures to minimise any possible damage by evacuating people from the probable danger zone. Nonetheless, when the main eruption occurred on 18 May, the effects were devastating (Fig. 1.1).

A large bulge which had developed on the

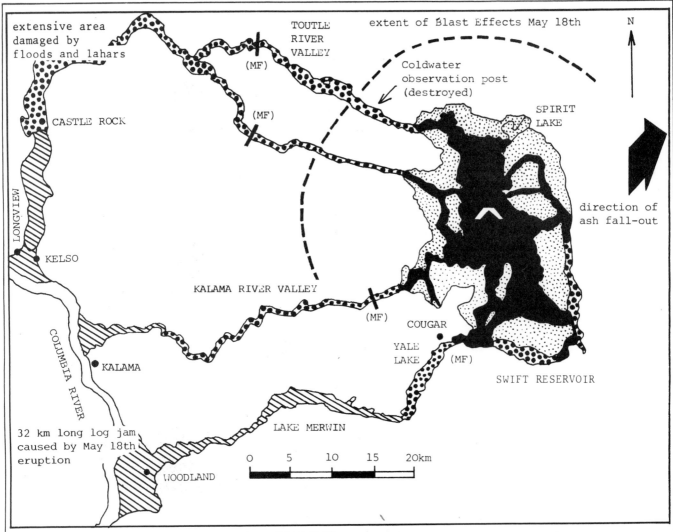

ZONES OF POTENTIAL HAZARDS AROUND MOUNT ST HELENS

- ⬛ pyroclastic surges, lava flows, mud flows and floods
- ⣿ lahars (mud flows) and flows
- (MF) approximate limit of thick lahars (mud flows) of May 18
- ⣿ ash clouds associated with pyroclastic surges
- ⧄ floods
- ╲ dam

Fig. 1.2 The effects of the eruption of Mount St Helens, May 18 1980.

flank of the volcano collapsed, producing a massive landslide. Within seconds, the gas-filled magma which had been 'bottled-up' beneath was blasted into tiny fragments and dust particles, which rolled down the mountainside as an incandescent cloud, travelling at speeds of up to 180 km h^{-1}. A few minutes later, an ash 'fountain' shot up some 19 km into the sky, destined to fall out afterwards over a vast area, covering homes, shops and streets with a thick grey blanket. Three days later the ash cloud reached the Atlantic coast over 4000 km away and in another 17 days it had completely circled the globe at an altitude of 10 km.

Two hours after the eruption, snow and ice from the summit, melted by the heat and mixed with the ash, formed mudflows known as lahars. These careered down neighbouring valleys at speeds of 80 km h^{-1}. Only the driver of a fast car on a clear highway would have been able to escape such forces (Fig. 1.2). Virtually no lava was emitted from Mount St Helens.

The eruption was estimated to have had the power of 10 megatons of TNT and it left 61 people dead or missing and caused an

estimated £1,000,000,000 damage to property. This was in spite of the precautions taken to clear the area, precautions which, before the event, were considered by many to be unduly strict.

Yet, it has been estimated that the eruption was only 2% as powerful as the 1815 eruption

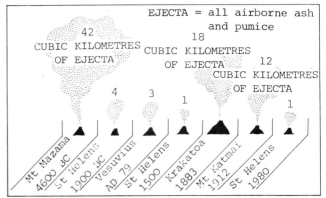

EJECTA = all airborne ash and pumice

42 CUBIC KILOMETRES OF EJECTA

18 CUBIC KILOMETRES OF EJECTA

12 CUBIC KILOMETRES OF EJECTA

Mt Mazama 4600 BC — 42
St Helens 1900 BC — 4
Vesuvius AD 79 — 3
St Helens 1500 — 1
Krakatoa 1883 — 18
Mt Katmai 1912 — 12
St Helens 1980 — 1

Fig. 1.3. Comparative output of ash and pumice from several volcanoes.

of Tambora in the East Indies. Many other volcanoes have produced far more airborne ash and pumice than Mount St Helens in 1980, as

Fig. 1.3 shows. Even the volcano itself produced four times as much material in 1900 BC than it did in 1980 AD.

Why should one volcanic eruption pass unnoticed by all except a few sailors and another give rise to the declaration of an emergency in the United States?

Why do some volcanoes quietly produce millions of cubic metres of lava with scarcely any ash, whilst others violently expel equally huge quantities of ash and only negligible amounts of lava?

In many cases, volcanoes erupt in different ways at different times of their existence.

Why is this? Whilst a volcano is forming on the earth's surface, what is happening at depth beneath it? Can we find evidence of former deep-seated igneous activity, now exposed to view by uplift and erosion? If so, can we establish any link with the magmas which we can still observe in active volcanoes?

These and a host of other questions form some of the problems which igneous petrologists set out to solve. In endeavouring to answer some of them we shall range across quite a wide spectrum of ideas, all of which aid our understanding of the igneous rocks and the processes which created them.

Magmas and their Characteristics

The eruption that stranded the ship at sea took place underwater and apart from collecting pumice, there were few further observations which could be made. However, this type of eruption is typical of many volcanoes of the ocean floor, some of which have built themselves up above sea level to form volcanic islands. Volcanoes of this quieter type are mostly found on oceanic ridges such as the Mid-Atlantic Ridge. The island of Tristan da Cunha, which lies adjacent to the Ridge, provides us with a good example. Its eruptive history is similar to the underwater activity which affected the ship.

Tristan da Cunha erupted unexpectedly in 1961 and its population had to be hastily evacuated. There was no loss of life, since the main product of the eruption was a thick flow of mobile lava, although this unfortunately

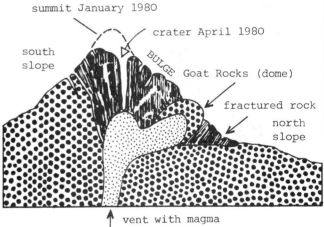

Fig. 1.5 Diagrammatic section of Mount St Helens shortly before the main eruption of May 18 1980.

obliterated the crawfish canning factory which was the main source of income. It also filled up the only bay on the island where boats could land and extended the coastline a further 500 m. After the main eruption, a team of scientists landed on Tristan da Cunha and studied it as intensively as some of their ill-fated colleagues were to do nearly 20 years later on Mount St Helens.

Careful mapping of each volcano has enabled us to draw sections of their likely structure (Figs. 1.4 and 1.5). The Mount St Helens section shows the volcano a few minutes before the huge bulge slid down the mountainside, releasing the pressure on the magma beneath, which was blasted out as gas-laden dust and ash. Most of the previously existing structure of the volcano was composed of congealed magma and debris. None of the old lava had flowed very far, since any which did not explode was exuded into a high-sided dome before it solidified.

By contrast, much of Tristan da Cunha, and presumably the submarine mound on which it rests, is composed of lava, interspersed with

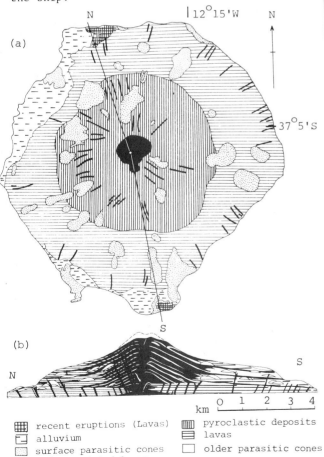

recent eruptions (Lavas)
alluvium
surface parasitic cones and associated lavas
peak cinder cone
pyroclastic deposits
lavas
older parasitic cones
intrusive rocks

Fig. 1.4 (a) Simplified map of Tristan da Cunha showing various features. (b) Diagrammatic cross-section of the Tristan da Cunha central-vent volcano.

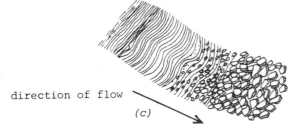

direction of flow

(c)

Fig. 1.6 The relationship between the structure of a lava and its position in the lava flow. (a) Ropy lava (pahoehoe), Kilauea, March 1974. (b) Blocky lava (aa), Kilauea, October 1977. (c) Lava with both pahoehoe and aa surfaces

ash bands. Individual flows can be traced for considerable distances and it is evident that the lava was far less viscous (i.e. more mobile) than in the case of Mount St Helens.

In seeking reasons for the differences, a first reaction might be to say that the sea water has a moderating influence on the behaviour of the oceanic volcanoes which is not present on land. However, volcanoes which normally behave in this mild manner are mostly limited to the oceanic ridges. Volcanoes of the island arcs, such as the East Indies, have been known to erupt with even more ferocity than Mount St Helens. Of the volcanoes named in Fig. 1.3, Mount Mazama (now Crater Lake, Oregon) and Mount St Helens are over 130 km inland, Vesuvius and Katmai are within a few kilometres of the coast and Krakatoa is the remains of a volcanic island off the coast of Sumatra. Proximity to water does not seem to provide the answer.

We have already indicated that the behaviour of a magma depends upon its viscosity. Can we determine the factors which affect this property? In the case of the more predictable volcanoes, much can be done by direct, close observation, experiment and sampling of all the products of vulcanicity, ranging from the molten lava itself to the gases which are given off. Temperatures can be measured by thermocouple or optical pyrometer. Observatories such as that of the US Geological Survey on Hawaii have long contributed a great deal to our knowledge. Collecting similar information from a volcano as unpredictable as Mount St Helens is more hazardous and indeed one geologist was killed whilst trying to observe the eruption in 1980. Nonetheless, many valuable data have been gathered, and further experiments have been done in the laboratory on remelted lavas and artificial melts.

The results of such work show that there are three main factors which control the viscosity and hence the likely eruptive properties of magmas. The first of these is temperature. This is illustrated in Fig. 1.6 which shows two markedly different types of lava flow on Kilauea, in Hawaii. The lava in Fig. 1.6a has a 'ropy' surface, known to the Hawaiians as pahoehoe, whilst Fig. 1.6b shows a crumbly, blocky lava flow, known locally as aa. It might be thought that the two are quite unrelated, but there are many cases where pahoehoe passes downstream into aa without any significant differences in the chemical composition of the two forms (Fig. 1.6c). The more viscous downstream part of the flow would clearly have been at a lower temperature, than the mobile upstream part. Other factors, however, such as the internal turbulence of the flow must also be important, or pahoehoe would never solidify but would always change to aa as it cooled.

Such differences in temperature at different stages of a single lava flow are reflected on a bigger scale when one compares the temperature at which an ocean ridge lava erupts with that of the more explosive lavas, when they reach the surface. The former is normally between 1050°C and 1200°C whilst the latter may be as low as 700°C to 800°C.

Another factor is the gas content of the magma. Considerable quantities of gases are produced during volcanic activity, notably water vapour, carbon dioxide, sulphur dioxide and trioxide, nitrogen, carbon monoxide, hydrogen, argon, sulphur and chlorine. Most of these gases are held in solution in the magma until it erupts, whereupon the release of pressure allows them to escape into the atmosphere. Experiment has shown that a rise in dissolved gas content lowers the viscosity and renders a lava more mobile.

The third main factor is the silica content of the magmas. The vast majority of igneous rocks are comprised mainly of silicate minerals, i.e. minerals which contain the elements silicon and oxygen as part of their chemical structure. Some rocks contain 'free' silica in the form of quartz but the biggest part of most rocks is composed of silica in combination with other elements such as calcium, sodium, potassium, aluminium, iron and magnesium. Whatever the actual mineral content, it is not unduly difficult to crush a lava sample and measure the proportions of each chemical component. It is then that we find a distinct correlation between the silica content of the lava and its behaviour when molten. Generally, lavas of high silica content also have a high viscosity; in other words, they are 'sticky' and will not readily flow.

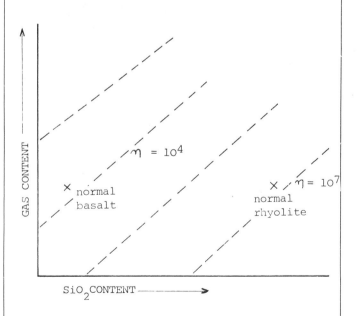

Fig. 1.7 *The relation of silicate melt viscosity to SiO_2 and volatile content.*

Figure 1.7 summarises, in semi-quantitative form, the influence of gas content and silica content on magma viscosity. Viscosity (η) is traditionally measured in poises (1 poise = 10^{-5} newton seconds per square metre). The graph shows the normal positions of two extremes of lava type, roughly at their temperature of crystallisation at atmospheric pressure. These are basalt (a mobile lava of viscosity 10^4 poise) and rhyolite (a sticky lava of viscosity 10^7 poise).

157

2
The Classification of Igneous Rocks

The Need for a Scheme of Classification

Such considerations of viscosity alone may help to explain the differences between volcanoes, but we also need to enquire why there are different magma types in the first place and what the ultimate origin of magmas may be. This involves studying phenomena at greater depth within the earth and considering how other igneous rocks are related to the lavas. Figure 2.1 shows clearly that there often is a connection between magmas

Fig. 2.2 Coire na Creiche, Cuillin Hills, Skye. The hills in the background are composed of gabbro (coarse-grained). In the foreground is an inclined sheet of medium-grained dolerite cutting the gabbro.

Fig. 2.1 Dyke cutting through lavas and ashes, Montagu Island, South Sandwich Group.

erupted as lavas and those which consolidate below ground. The photograph, taken on an island in a chain of recently active volcanoes in the South Atlantic, shows a sequence of lava and ashes which has been partly eroded away to reveal a wall-like dyke. This is made of igneous rock which is chemically similar to the lavas, but which differs from them in that the crystals had longer to form and are therefore larger.

Figure 2.2 is of part of the Cuillin Hills, Isle of Skye, which are composed of yet another igneous rock of the same composition but of coarser grain size still. No one can observe such intrusive rock being formed. It is believed to have cooled considerably more deeply in the crust than the dyke, so it cannot be so easily linked with surface volcanic activity. There are, however, so many chemical and mineralogical similarities

to lavas being erupted today that it is natural to seek some links. Obviously, measurements of magma temperature, gas content and rate of flow are quite impossible once the rock is formed, so we must turn to other criteria. In doing so, we hope to develop a method of classification which, if it is to be of any value, must cover the whole range of igneous rocks, from lavas to deep-seated intrusions, from ancient to modern. Any system must be flexible enough to fulfil a wide range of needs.

Geological investigations may include the examination of hand specimens in the field, detailed study of thin slices of rocks beneath the microscope, chemical analyses of crushed rocks or of extracted minerals, as well as observations of actual volcanic processes as outlined above. Many of the products of igneous activity have economic uses, so here is another need for careful definitions. Students will also be painfully aware of the fact that they are often expected to identify odd lumps of rock straight from the drawer, without reference to any helpful aids such as knowledge of the field relationships, rock chemistry, or even thin sections for the microscope! Inevitably, any one system of classification is bound to be in the nature of a compromise, which will need adaptation by the various specialist interests, but the basic framework set out on the following pages has gained general acceptance among geologists.

Principles of the Classification of Igneous Rocks

Chemical composition

We have already indicated that many of the differences between lavas are due to varia-

tions in their chemistry, which provide one possible basis for classification. Although at one time laborious, it is now technically feasible for a well-equipped laboratory to measure, quite speedily, the chemical composition of a crushed sample of igneous rock. Results are usually expressed in terms of the oxides of the various elements present, not because they occur in this form but because this is how they were traditionally processed as part of the operation.

TABLE 2.1 COMPOSITIONS OF SOME IGNEOUS ROCKS (AVERAGE WEIGHT %)

COMPONENT	THOLEIITE BASALT	ALKALI BASALT	SPILITE	ANDESITE	TRACHYTE	RHYOLITE	PERIDOTITE
SiO_2	50.0	46.0	51.0	60.0	63.0	73.0	43.5
Al_2O_3	16.0	15.0	14.0	17.0	18.0	13.0	4.0
Fe_2O_3	2.0	4.0	3.0	2.0	2.5	0.5	2.5
FeO	7.0	8.0	9.0	4.0	1.5	1.5	10.0
MgO	8.0	9.0	4.5	3.5	0.5	0.5	34.0
CaO	12.0	9.0	7.0	7.0	1.0	1.5	3.5
Na_2O	2.5	3.5	5.0	3.5	7.0	4.0	0.5
K_2O	0.5	1.5	1.0	1.5	5.0	4.0	0.3
TiO_2	1.5	3.5	3.5	0.5	0.5	0.5	1.0
H_2O	0.5	0.5	2.0	1.0	1.0	1.5	0.7

Table 2.1 gives the results of analyses of some typical igneous rocks. The importance of the silica content of a lava has already been stressed and it is this which has been chosen as the main basis of chemical classification of all the igneous rocks. Many years ago, arbitrary divisions were chosen and the spectrum of igneous rocks divided into four categories. The names of each of these were based upon a misconception of the actual properties of silica, but the names have remained, in spite of attempts to find better ones. The categories are shown in Table 2.2.

TABLE 2.2

PERCENTAGE OF SiO_2 in BULK CHEMICAL COMPOSITION	NAME OF CATEGORY
	ultrabasic
— 45 —	basic
— 52 —	intermediate
— 66 —	acid

The main advantage of this system is that it enables the geologist to relate the solidified rock to its probable magmatic source. For example, the lavas of oceanic ridges look very similar to those of ocean fracture zones and yet there are differences in their chemistry which reflect their different origins (i.e. tholeiite-basalt and alkali-basalt in Table 2.1). On the other hand the lavas, dyke rock and deep seated intrusive already referred to

look very different from each other and yet analyses of them show that they are chemically almost identical.

The main disadvantage is one of inconvenience. We need names to apply to specimens as we find them, without having to await the laboratory's report. Another complication is the rather arbitrary way in which the divisions were drawn up in the first place. After many thousands of analyses, we know that the division at 66% SiO_2 coincides with a minimum in the frequency curve of all the rocks analysed, so it is quite well chosen (Fig. 2.3). The other two dividing lines are badly selected, however, the 52% line actually coinciding with a peak of abundance of rock types!

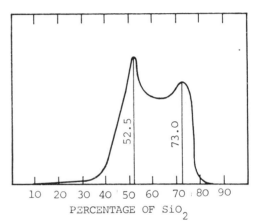

PERCENTAGE OF SiO_2

Fig. 2.3 The frequency distribution of silica percentage in analyses of igneous rocks.

In spite of these problems, the concept of a range of rocks from ultra-basic to acid is of considerable value, although in practice the label is more frequently derived from other criteria related only approximately to the silica percentage. Variation in other chemical components is also of importance, yet it is not expressed by this method.

Colour

We have seen that rocks may only be grouped on the basis of their chemistry after laboratory analysis. By contrast, one of the most obvious properties of a rock which one can easily record is its colour. Igneous rocks range from pale grey or white through to black. This often, although not always, reflects significant differences in rock chemistry or mineral content and it may be used as a rough basis for classification. The terms used are based on Greek words and are as follows:

light coloured - leucocratic 0-30% dark minerals
medium coloured - mesocratic 30-60% dark minerals
dark coloured - melanocratic over 60% dark minerals

The method provides a useful basis for rough divisions but it reveals nothing about the genesis of the rock. Also, whilst most of the melanocratic rocks are basic or ultra-basic types there are some acid rocks which are black. A good example is the shiny black volcanic glass, obsidian (Fig. 3.7b).

More recently, other terms have entered common usage in an attempt to describe concisely the dominant types of minerals present in the rock. Thus 'felsic' describes a rock in which feldspar and quartz (silica) are major constituents. 'Mafic' is used for rocks which contain some feldspars but which are rich in the ferromagnesian minerals (i.e. magnesium and iron (Fe) bearing). 'Ultra-mafic' labels a rock which is almost completely made of ferromagnesian minerals alone. The felsic minerals are mostly light-coloured and the mafic ones dark, so for most practical purposes the terms have come to be applied on the basis of the over-all colour of the rock, and are often loosely regarded as alternatives to 'leucocratic', 'melanocratic', etc.

Mineral content

Most students have seen and handled good specimens of a range of minerals. In doing so, they have learned that most minerals have clearly defined properties which enable them to be identified. The majority of igneous rocks are composed of minerals which have crystallised tightly packed together. Identification of them is not so easy as it is for the individual mineral specimens but with a little practice it can be done, either in the hand specimen, or with a thin-section of the rock beneath the petrological micro-scope, or both. The relative abundance of each mineral can also be estimated. Because of the wide range of chemical constituents comprising rocks from acid to ultrabasic, there is an equally diverse number of minerals which may be present. Here then is a potentially useful way of establishing divisions, both in the field and in the laboratory.

Over 3,000 minerals are known, but, fortunately for the petrologist (person who studies rocks), most of these are uncommon! In practice, the majority of igneous rocks can be described in terms of a dozen or so which are usually referred to as the rock-forming minerals. The main groups of minerals found are quartz, the feldspars, the micas, ferromagnesian minerals such as olivines, pyroxenes and amphiboles, and the iron ores, notably magnetite.

The table in Fig. 2.4 has been constructed empirically from the results of many thousands of mineralogical analyses of igneous rocks. It shows the variation in the proportion of each of the main rock-forming minerals and how this is related approxi-mately to the chemical division described before. The ferromagnesian minerals are listed in order of abundance.

The minerals listed are usually referred to as the essential minerals and it can be seen that there is only a handful of such minerals in each category, e.g. basic rocks consist essentially of plagioclase and a pyroxene such as augite. Other minerals are commonly present, but usually in smaller quantities and these are known as accessory minerals. Examples would include sphene, apatite or zircon in acid rocks. In the

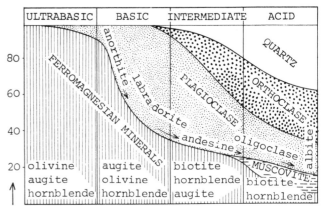

APPROXIMATE PERCENTAGE BY VOLUME OF FERROMAGNESIAN MINERALS

Fig. 2.4 *The chief minerals of igneous rocks.*

basic category, olivine, although listed in the table, is often regarded as an accessory mineral. When appropriate, its presence is indicated by a hyphen, including the word as a prefix to the rock name, e.g. olivine-basalt.

In identifying igneous rocks from their mineral content, it is usual to approach them systematically:

1. Is quartz present and if so in what proportion? The table shows that acid rocks can contain over 20% quartz, intermediate ones between 20% and a few percent and basic ones very little. When quartz does occur in a basic rock it is usually regarded as an accessory. IT IS VITAL NOT TO CONFUSE THE PERCENTAGE OF THE MINERAL QUARTZ IN THE ROCK WITH THE PERCENTAGE OF SILICA IN ITS BULK CHEMISTRY. They both share the formula SiO_2, yet in the chemical analysis the SiO_2 percent-age is the total of all the silica, occurring both in the free state and in combination in other silicate minerals. Thus, a basalt may have a SiO_2 content of 50% and yet have no 'free' quartz.

2. What is the feldspar content and which feldspars are present? The proportion of the rock which is composed of feldspar can usually be estimated in the hand specimen, as can the distinction between potash feldspar (orthoclase etc.) and the plagioclases. The determination of the variety of plagioclase, however, is normally only possible with the aid of a petrological microscope. A slice of rock is ground down until it is so thin that most of its minerals are transparent. This is examined beneath a microscope equipped with polarised light and a rotating stage. The optical properties of minerals in such thin sections are quite characteristic and it is not difficult to identify them to the level required by Fig. 2.4. It is important to be able to work in this detail, since the classification partly depends on the type of plagioclase, ranging from the calcium-rich types (anorthitic) in the basic rocks to the sodium-rich (albitic) in the acid ones. Potash feldspar and the albite variety of plagioclase feldspar are commonly referred to as the alkali feldspars.

3. Are micas present and if so which ones?

The white mica, muscovite, is most often found in the acid rocks and helps to give them their lighter colour. Biotite is often present in acid rocks but is most abundant in the intermediate ones.

4. Which ferromagnesian minerals are present and what are the proportions? The ultra-basic rocks are mainly composed of the darker, denser ferromagnesian minerals and it is perhaps in this category that distinguishing between them is of greatest importance. The table (Fig. 2.4) shows the most common ferromagnesian minerals in each group, and it also shows that some minerals are mutually exclusive. For example, it would be very unusual to find both olivine and quartz in the same rock; if sufficient silica is present to form free quartz, then olivine crystals would not have survived for long in the magma, but would have reacted with the excess silica to form an amphibole or a pyroxene. (See p.20)

It must be evident from the above detail that the mineral content of a rock provides a powerful basis for its classfication. Nonetheless, there are some drawbacks. It does not distinguish between rocks formed at different depths, thus fine-grained rocks often have to await sectioning and microscope examination before a positive identification can be made. It is also easy to ignore possible genetic relationships between rocks by putting them in neat categories. For example, small quantities of acid lavas are found in association with basic ones and may represent later derivations from the same magma.

Texture

By the word <u>texture</u> we mean the grain size of the rock and the relationship between neighbouring mineral constituents. We shall examine some particular textures later but it is sufficient here to note the relationship between the size of the crystals in an igneous rock and the time which it took to cool. The concept will be familiar to anyone who has seen salol crystallise at different rates or has tried to grow large copper sulphate crystals from aqueous solution. Generally, a slower rate of cooling produces bigger crystals.

In geological terms, slower cooling rates are achieved deeper in the crust than at the surface, so the use of grain size, as a criterion for classification, will usually reflect the geological environment in which the rock crystallised. In the past, such principles were probably taken a little too far, and the terms 'volcanic', 'hypabyssal' and 'plutonic' were applied to the hand-specimen to indicate, respectively, whether the rock was produced at the surface, at moderate depths, or deep in the crust. Such ideas are most useful in the field, where larger scale structure may also be observed, but it is a little risky in the hand specimen. For example, not all fine-grained basalts are lavas; basalt quite frequently occurs as dykes, intruding other strata.

A Classification of Igneous Rocks

In practice, the most useful classification system is one which combines the mineral content of a rock with its texture. The approximate relationship between mineral content and chemical composition has already been shown in Fig. 2.4 and Table 2.2. In Table 2.3 which shows an outline classification of igneous rocks, texture has been plotted against the approximate chemical categories only, for the sake of simplicity.

Both Fig. 2.4 and Table 2.3 therefore need to be read together in order to relate mineral content and texture. For example, a basalt is a fine-grained basic rock and gabbro is a coarse-grained one, but they are chemically similar and contain the same essential minerals, namely calcic plagioclase and pyroxene.

Petrologists have set up a host of names for varieties of igneous rocks but in Table 2.3 we have used the rock names in their general sense. For example, the name 'granite' covers a family or clan of coarse-grained acid rocks.

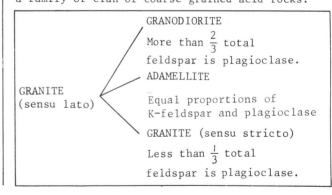

TABLE 2.3 A CLASSIFICATION SCHEME FOR IGNEOUS ROCKS

"ULTRABASIC"	"BASIC"	"INTERMEDIATE"		"ACID"	VOLCANIC & MINOR INTRUSIVE ROCKS
		plagioclase> K feldspar	K feldspar> Plagioclase		
(rare)	basalts	andesites	trachytes	rhyolites obsidians	fine-grained or glassy
(rare)	dolerites	microdiorites	microsyenites	microgranites	medium to fine-grained
peridotites serpentinites	gabbros	diorites	syenites	granites	PLUTONIC ROCKS coarse to medium- grained

The alternative names in the fine-grained acid rock categories reflect marked textural differences, rather than mineralogical ones. The intermediate rocks are divided into two groups because of significant differences in mineralogy, which have long been recognised by the use of these names. In the trachytes and syenites, the proportion of alkali feldspars (Na and K rich) is relatively high. The andesites and diorites are proportionately richer in the calcium feldspars.

It must be stressed again that such a fitting of rocks into pigeonholes is a convenient but rather artificial way of seeking order in the natural world. Many geologists prefer alternative systems, but all would agree that whatever system is chosen, rocks do not fall into neat slots but form parts of a continuously varying spectrum.

Since so much of our understanding of igneous rocks depends upon microscope examination of thin sections, a series of drawings of such sections follows (Fig. 2.5). The drawings are in the same order as in Table 2.3, for ease of reference. The minerals have been shown by partly stylised symbols, which approximate more closely to the view seen in plane polarised light than to crossed polars (i.e. two pieces of Polaroid with their polarisation axes at right angles).

Fig. 2.5 A selection of igneous rocks in thin section.

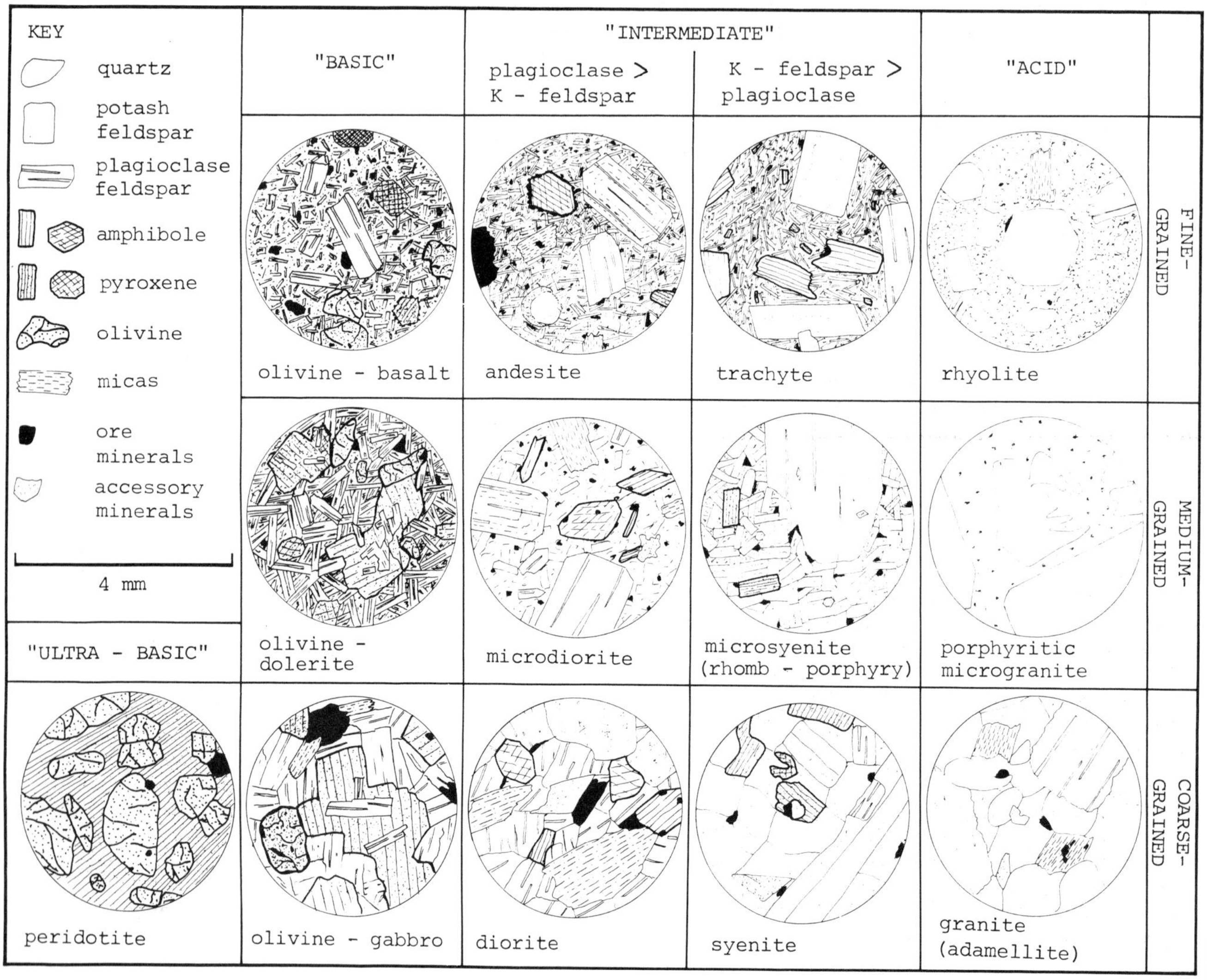

3
Determining the History of an Igneous Rock

We have discussed the classification of igneous rocks at some length, since more is at stake than simply assigning a name to a given piece of rock. The use of the correct name tends to make most geologists think not just of a lump of rock, but also of its origin and its significance in global tectonic terms. However, there is more information to be considered which can add a great deal to our knowledge of the origin of an igneous rock. Broadly, this falls into three categories: laboratory work on silicate melts; detailed examination of rock textures; observations of the structure and field relationships of igneous rock masses.

Silicate Melts in the Laboratory

Laboratory experiments have been carried out over the last 70 years or so, since the pioneer work of N L Bowen and his associates in the USA. In view of the very high temperatures and pressures involved, such work requires highly specialised equipment and skills, but it is now possible to imitate natural processes by remelting existing rocks, or by making synthetic ones. We can thus gain vital information regarding the temperatures and pressures at which rocks may begin to melt to produce magmas. It is also possible to predict the cooling history of a magma, which may then be compared with the field and laboratory evidence from the rocks themselves. Some of the most important results concern the effects on melting points of mixing several components or of introducing water vapour into the system.

Figure 2.4 showing mineral content indicates that most igneous rocks consist of two or more essential minerals. Of course, accessory minerals are also usually present. Many experiments have been conducted with more than two components present, but the results are rather difficult to understand in graphic form and we shall limit ourselves to melts with only two components. A simple analogy of the principles involved is the way in which salt lowers the freezing point of water. We shall consider three examples.

The diopside-anorthite system

Diopside and anorthite are two relatively simple silicate minerals. Diopside ($CaMg Si_2O_6$) is a pyroxene and anorthite ($CaAl_2 Si_2O_8$) is an end-member of the plagioclase feldspars. In a way, this mixture may be regarded as a simple version of a basic rock such as basalt. The graph (Fig. 3.1) is derived from the experimental work carried out at atmospheric pressure, and it shows the effect which each component has on the other. Thus, pure diopside melts at 1391°C and pure anorthite at 1660°C. The outcome of mixing the two is to lower the melting point, the amount depending on the proportions of each component present. The lowest melting point on the graph is at 1270°C, corresponding to a composition of 58% diopside and 42% anorthite. This point is known as the <u>eutectic</u>.

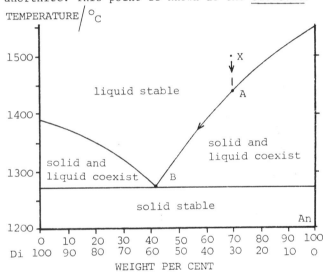

Fig. 3.1 *Equilibrium diagram of the system anorthite (An) diopside (Di).*

The graph may be used to determine the crystallisation history of a melt as follows. Assume a melt of composition 30% diopside, 70% anorthite at a temperature of 1500°C (point X on the graph). As the temperature falls, there is at first no change in the composition of the melt, nor do any crystals form, since only liquid is stable above the line. When the temperature reaches 1440°C, anorthite begins to crystallise, but not diopside. Because the components of anorthite are being removed from the liquid by crystallisation, the composition of the liquid will change along the line AB. Once the temperature has fallen to 1270°C and the composition has reached the eutectic proportions ($Di_{58} An_{42}$) then diopside will crystallise. After this, the two minerals crystallise together in the eutectic proportions, and the temperature remains the same until the liquid is all used up and the mass has set solid.

Reheating the solid material is the exact reverse of the above. The first liquid appears at 1270°C and the temperature does not increase until all the diopside and most of the anorthite has melted.

The graph relates to a 'dry' system. The addition of water vapour, which is very often present in real melts, lowers the melting points and moves the eutectic to the right. For example, at a water vapour pressure of 10^9 Pa (10 kilobars), the melting point of pure diopside drops to 1280°C and that of anorthite to about 1130°C. The eutectic is then at 1020°C. The applications of such studies to igneous petrology are quite far-reaching and some will be dealt with in later sections.

Crystallisation of the olivines

The pale green mineral which most of us know as olivine is in fact one member of a whole family of such minerals. The olivines are silicates of iron (Fe^{2+}) and/or magnesium (Mg^{2+}), e.g. $MgFeSiO_4$. The ions of magnesium and iron are each of valency 2 and their dimensions are comparable. (The ionic radii are: $Fe^{2+} = 0.83Å$; $Mg^{2+} = 0.78Å$.) This means that Mg^{2+} and Fe^{2+} are easily inter-changeable within the well-ordered lattice structure of the mineral and so a continuous range of composition is possible, from an all-magnesium variety of olivine (Mg_2SiO_4 - forsterite) to an all-iron variety (Fe_2SiO_4 - fayalite). This phenomenon of interchange-ability of ions is known as isomorphous substitution (isomorphous means 'same shape', i.e. there is no change in the crystal lattice structure throughout the series). The series is also known as a 'solid solution series', since it appears as though one component is 'dissolved' in the other on the molecular scale. Although the chemistry of the olivines is rather simpler than that of the diopside-anorthite system, the melting-point curve is perhaps a little more difficult to follow.

The diagram (Fig. 3.2) shows the cooling curves for the olivine family from forsterite to fayalite. The melting point of pure forsterite is 1890°C and that of pure fayalite is 1205°C.

Consider a magma of composition 50% forsterite, 50% fayalite (i.e. $Fo_{50} Fa_{50}$) at a temperature of 1800°C, shown as point X. The magma will cool to 1650°C (Y) before any crystals appear. When they do, they will have the composition shown by the lower line at that temperature, i.e. point Z on the graph; a composition of $Fo_{80} Fa_{20}$. Such crystals will be in equilibrium with the melt. As the temperature drops, however, the composition of the melt changes along the upper line and that of newly-formed crystals changes along the lower line, in the direction of the arrows. This has some interesting implica-tions. Unless the early-formed crystals have been removed from the melt, perhaps by sinking to the bottom of the magma chamber, they will find themselves surrounded by a liquid with which they are no longer in equilibrium. Hence, fresh mineral growth around the margins of such crystals will be more iron-rich than the original core of the crystal and it will appear 'zoned'. In a rock

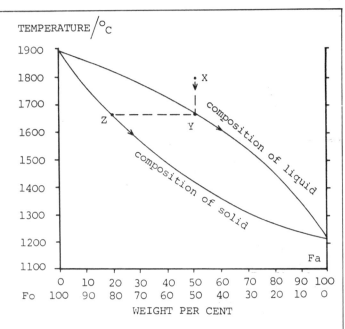

Fig. 3.2 Equilibrium of the system forsterite (Fo)-fayalite (Fa).

which contains larger, early-formed crystals (phenocrysts) set in a finer-grained ground-mass, the large crystals may be markedly richer in magnesium than the later, smaller ones. If cooling is slow enough, the whole crystal may be altered by such reaction.

Crystallisation of the plagioclase feldspars

The plagioclase feldspars also form a continuous series of isomorphous substitution (solid solution) between albite ($NaAlSi_3O_8$) and anorthite ($CaAl_2Si_2O_8$).

The crystallisation diagram is shown in Fig. 3.3 as an exercise for you. The answers are given at the end of the Unit. Again, the influence of water vapour (not shown) serves to lower the melting points of the whole system.

Crystallisation of magmas

Real magmas usually contain more than the two components considered in each of the above studies; their crystallisation curves become progressively more complex and they are not dealt with in detail here. Of vital interest, however, is the behaviour of real magmas at different depths in the earth, and much of the experimental work has been directed at studying the effects of pressure as well as temperature. Many magmas also contain appreciable quantities of water vapour and this has a marked influence on their crystallisation history, as already indicated for the diopside-anorthite system.

Basic magmas

Figure 3.4 shows a simplified melting-point curve for basic magma, where temperature is plotted against pressure. The pressure can also be roughly equated to depth of burial in the earth. The curve is that for 'dry' basic magma. Although much water vapour is exhaled during volcanic eruptions, it forms a relatively minor constituent of basic

Study the graph below which shows the crystallisation of plagioclase feldspar under equilibrium conditions. A liquid melt of mixed composition cools to temperature X ($1500°C$) at which crystallisation begins. The crystals which form at this temperature may be represented by point Y on the lower curve and have a composition Z (almost pure anorthite). As the temperature continues to fall, the composition of the liquid melt changes in accordance with the upper curve on the graph. The composition of the crystals also changes - both for those already formed, and for those that will form at the lower temperature. The liquid melt thus remains in equilibrium with its crystals.

Figure A, which refers to part (c) of the question, shows a zoned crystal of plagioclase feldspar in which the composition of each layer has *not* changed during crystallisation. Answer the following questions.

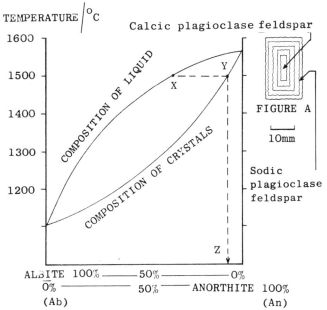

(a) Account for the change in composition of plagioclase feldspar which occurs during crystallisation from a liquid of composition X under equilibrium conditions (very slow cooling).

(b) What are the properties that enable ions to substitute for one another to permit the change from anorthite ($CaAl_2Si_2O_8$) to albite ($NaAlSi_3O_8$)?

(c) In some basic igneous rocks, zoned plagioclase crystals are formed, for example the crystal in Figure A. Explain, briefly, how the zoned crystal could have been formed.

Fig. 3.3 Exercise based on JMB A level paper.

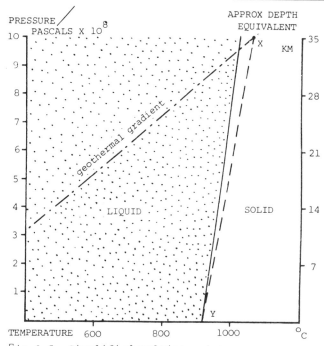

Fig 3.5 Simplified melting point curve for "dry" acid magma.

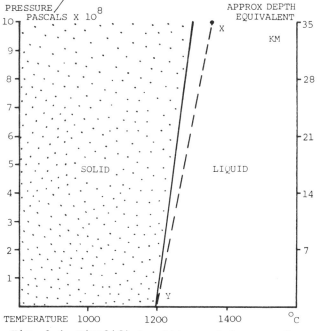

Fig. 3.4 Simplified melting point curve for "dry" basic magma.

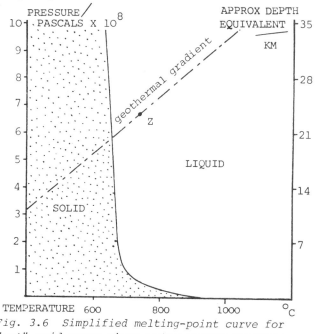

Fig. 3.6 Simplified melting-point curve for "wet" acid magma where water vapour pressure equals load pressure.

magmas and can be ignored for the present purposes. Assume for the sake of this exercise that a basic magma forms at 1350°C and 10^9 Pa pressure. As it rises through the earth, assume that it cools along the line XY (remember that rising in the earth means moving <u>down</u> on the graph). The line XY intersects the melting point curve at the earth's surface, indicating that the magma will still just be in liquid form and capable of erupting as lava.

Although this presents an oversimplified view, it is not too far from the truth and helps to explain why basaltic lavas are so common.

Acid magmas

Figures 3.5 and 3.6 illustrate melting point curves for acid magmas. Figure 3.5 is for 'dry' magmas, whilst Fig. 3.6 is one of many which could be drawn for an acid magma containing water vapour. In this case, the water vapour pressure is assumed to be equal to the load pressure from the overlying crust. The difference between the two graphs is remarkable, the effect of the water vapour being to lower the melting point by as much as 400°C.

Many acid magmas originate by the melting, or partial melting, of the rocks of the lower crust, consequently the 'normal' geothermal gradient within the crust has been added to the diagrams.

It can be seen that for a melt originating at point X, rising in the crust and cooling at the rate of say 4°C per km (XY) it is capable of reaching the surface and erupting as lava.

Now try the same exercise for the 'wet' acid magma, assuming a starting point of Z, and the same rate of cooling. How far will it rise before it begins to crystallise? (See Appendix.)

In practice, it would seem that 'wet' acid magmas are more common than dry ones, perhaps because of the pore-water contained in the crustal rocks from which they are formed. Certainly, acid lavas (rhyolite, obsidian etc) are less voluminous than granite masses. We shall later consider the evidence for the cooling of both basic and acid magmas with some case studies. First, we shall examine the importance of the textures of igneous rocks as an aid to understanding their origin.

Textures of Igneous Rocks

General textures

A considerable amount may be learned about a rock from its texture. The importance of grain size in classification and as an approximate guide to the depth at which the rock crystallised, has already been outlined. However, there exists a wide variety of detailed textures, many of which have frightening Greek names and some of which are not easily understood. Some textures are clearly visible to the naked eye in the hand specimen, some require the assistance of a good hand-lens, and many can only be studied in thin section beneath the microscope. At the risk of divorcing rock textures from the rest of the data about the rock, it may be helpful at this stage to catalogue the

various possibilities, with some indication about their origins. In some cases, a direct link will be observed with the crystallisation diagrams of the last section.

Table 3.1 outlines the main characteristics of each texture but the drawings or photographs are shown separately (Fig. 3.7).

You should try to match each drawing or photograph to the appropriate description. Inevitably, the drawings show textures taken out of context. When you have checked answers (at the end of the Unit) it would be a useful exercise to turn back to the thin-section drawings in Fig. 2.5 and see how many textures can be identified in the whole rock and what can be inferred about its origin.

(a)

(b)

(c)

(d) 2 mm   (e)   (f) 1 mm (g)   0.5 mm   (h) 2 mm

    0.2 mm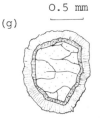

III   I

Match these photographs and sketches to the descriptions of igneous texture in Table 3.1. The key to the minerals in the sketches is the same as in Fig. 2.6

Fig. 3.7 Textures of igneous rocks in the hand specimen and in thin section.

TABLE 3.1 TEXTURES IN IGNEOUS ROCKS

TEXTURES	EXAMPLES	DESCRIPTION	A POSSIBLE INTERPRETATION	
Glassy texture	Devitrified obsidian	Usually black and shiny in hand specimen, with conchoidal fracture. Ancient volcanic glasses often have tiny crystals arranged radially in 'spherules', or along curving 'perlitic' cracks.	Result from supercooling of magmas where cooling was too rapid for crystallisation and the magma set as a glass. With time the glass crystallises, or devitrifies into tiny crystals.	
Grain size (not illustrated in Fig 3.7)	Coarse grained	Crystals of the ground mass are easily seen and mostly identifiable with the naked eye.	Crystals had plenty of time to grow around a limited number of nucleii. Usually typical of plutonic rocks.	COOLING TIME several millions of years
	Medium grained	Crystals of the ground mass seen with the naked eye but hand lens needed for identification.	Crystals formed more quickly around a greater number of nucleii. Usually typical of minor intrusions.	
	Fine grained	Crystals of the ground mass not distinguishable with the naked eye and not identifiable with a hand lens.	Crystals formed very quickly around many nucleii. Typical of lavas and chilled margins of minor intrusions	↓ perhaps less than a year
Grain shape	Euhedral	Well-formed crystals showing perfect or near-perfect crystal form.	Usually the first crystals to form in the magma and therefore unrestricted.	
	Subhedral	Grains show an imperfect but still recognisable crystal form.	Formed at a time intermediate between the early and late stages of crystallisation.	
	Anhedral	Grains show no regular crystal form.	Usually the last crystals to form, filling up gaps. Many rocks consist largely of equidimensional anhedral or subhedral crystals. Their texture is referred to as granular.	
Intergrowth textures	Granophyric	The rock consists of quartz and feldspar grown together in curious embayed crystal shapes. Associated with medium to fine-grained rocks.	Probably results from crystallisation at or near the eutectic point.	
	Graphic	Another quartz-feldspar intergrowth found in coarse or very coarse-grained rocks. Quartz crystals grow in angular sheets looking like ancient cuneiform writing. Ratio of quartz to feldspar usually 30 : 70	*Try to decide for yourself.**	
Reaction textures	Zoned crystals	Common in plagioclase feldspars. Crystals show concentric 'rings', the central portion generally being more calcium-rich than the outer zones which are relatively sodium-rich.	Reaction has taken place between crystal and liquid in a continuous solid solution series under relatively quick cooling. There is not time for the whole of the early-formed crystal to react with the changed composition of the melt.	
	Corona structure	This appears as a type of zoning, except that the zones are composed of different minerals, e.g. olivine in the core, with a rim of pyroxene, followed by an outer rim of amphibole.	Also results from reaction between early-formed crystals and the melt as its composition changes, so that the two are no longer in equilibrium. Occurs in minerals forming discontinuous reaction series, i.e. different minerals develop rather than variations of the same mineral.	
Some other textures	Ophitic texture (a type of poikilitic texture)	Small euhedral plagioclase crystals are enclosed by one large augite crystal. Most common in dolerites. Other pairs of minerals with similar relationships may occur. The general name for such pairs is poikilitic texture.	*Try to decide for yourself.**	
	Porphyritic texture	Large crystals, usually euhedral forms, are set in finer-grained groundmass. The large crystals are termed phenocrysts. Common in lavas and in some granites (e.g. Shap Granite).	*Assuming that crystal size is mostly controlled by rate of crystallisation, try to make your own decision about this one too.**	

* Answers to the illustrations and to the problems marked * will be found at the end of the unit.

Textures limited to particular groups of rocks

Pegmatites and aplites

Quite often, limited portions of plutonic rock masses and offshoot dykes exhibit unusually coarse-grained and unusually fine-grained rocks. The former are known as pegmatites (Fig. 3.8a) and they sometimes contain crystals of a metre or more across. A giant crystal of spodumene nearly 15m long was recorded in South Dakota. The most common types are associated with granite plutons and consist largely of quartz, feldspars and micas. However, they may also contain minerals rich in former volatiles such as boron and fluorine. Tourmaline is a good example.

The finer grained rocks are known as aplites and usually have a sugary appearance (Fig. 3.8b). They may contain similar minerals to pegmatites.

Both groups of rocks are regarded as the products of late-stage magmatic activity when the residual magmas were rich in water vapour and volatiles. The relationships between pegmatites and aplites remain rather speculative (Fig. 3.8c).

Lava textures

Lavas exhibit many of the textures listed in the table, but they are also characterised by other textural features which are not normally associated with intrusive rocks. The main ones are the following:

Vesicular texture

Vesicles are cavities in the lava produced by the escape of gases (Fig. 3.9a). Vesicular lavas are most typical of the tops of lava flows, where there is no pressure of over-lying material to keep the gas in solution. Sometimes a vesicular lava may occur near the base of a flow; in this case it has been carried down from the cooled top surface as the inside of the flow moved forward, rather like a tank-track.

Amygdaloidal texture

This Greek-looking term is derived from the word for almonds. It describes a lava where

(b)

(c)

Fig. 3.8 (a) A pegmatite with tourmaline, Porthleven, Cornwall. (b) A narrow aplite vein in granite, Shap Fell Cumbria. (c) Aplite and pegmatite veins in close association, Porthleven, Cornwall.

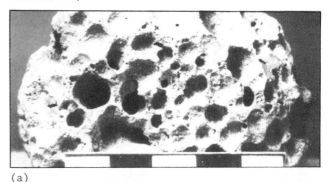
(a)

(b)

(c)

Fig. 3.9 (a) Vesicular lava, Mediterranean.` (b) Amygdaloidal lava with calcite filling almond-shaped vesicle. Locality unknown. (c) Amygdaloidal lava with two generations of zeolite crystals filling elongated vesicles. Locality unknown.

(a)

the vesicles have been filled in by later generations of crystals. In a weathered lava, these sometimes stand out as oval shapes, looking rather like sugared almonds (Fig. 3.9 b and c).

Flow-banded textures

Some of the more viscous lavas, usually those of acid or intermediate composition, exhibit a rough banding of their constituents (Fig. 3.10a). The bands are usually continuous for several centimetres and are frequently highly convoluted. A special case is trachytic texture, which is typical of trachytes and consists of a 'swirling' of tiny laths of potash feldspar crystals (Fig. 2.5).

(a)

(b)

(c)

Fig. 3.10 (a) Flow-banded rhyolite, Snowdon. (b) Ignimbrite, locality unknown. (c) Bedded tuff, Eifel Mountains, West Germany.

Textures of pyroclastic rocks

True flow-banding of viscous lavas is often confused with the texture of an ignimbrite, which is not produced by a lava flow at all. Ignimbrites (sometimes called welded tuffs) originate by the deposition and rapid 'welding' together of dust particles which travel down the slopes of a volcano in a great incandescent cloud, or nuée ardente. The hot dust cloud originates from the shattering of very viscous lava which has generally formed a dome-like mass within the surface layers of the volcano. The character-istic features of an ignimbrite are the discontinuous nature of the bands and the presence of crushed shards of glass looking like the letter 'Y' on its side (Fig. 3.10b).

The mode of origin of an ignimbrite is often referred to as a 'pyroclast flow' to dis-tinguish it from the other types of pyro-clastic activity, where the heat is not sufficient to produce the welded texture. These other pyroclastic rocks show great variation in particle size, ranging from badly sorted agglomerates through a variety of tuffs (Fig. 3.10c) to fine volcanic dust, which may settle to form a type of loess. The agglomerate in Fig. 3.11 is unusual in that it exhibits graded bedding, the fragments probably having fallen into water.

10 cm

Fig. 3.11 Agglomerate showing graded bedding.

In conclusion, comment on the texture of the rock in the photograph (Fig. 3.12). See the back of this Unit for an answer.

Fig. 3.12 Comment on the texture of this rock.

Major Basic Intrusions

Much can be learnt about the way magma behaves during crystallisation and about its origins by studying major intrusions. Most of the major intrusions fall conveniently into two broad groups, the basic and the granitic, using the terms in their broadest sense, and these will be considered separately.

There are a number of very large basic intrusions which can be quite informative about the way in which magma crystallises. In dealing with major basic intrusions it should not be forgotten that the rate of cooling is completely different from extrusive events and some major intrusions can take thousands or even millions of years to finally consolidate.

We have already outlined the significance of laboratory work by N L Bowen and others. In a sense, a large basic intrusion may be thought of as a gigantic natural laboratory where the results of processes acting over a long period of time may be preserved. Observations of reaction rims (Fig. 3.7) and the order of crystallisation from silicate melts (see earlier) have led to the formulation of an ideal sequence in which minerals could be expected to crystallise. This sequence has become known as 'Bowen's Reaction Series' and it consists of <u>two</u> lines of descent for

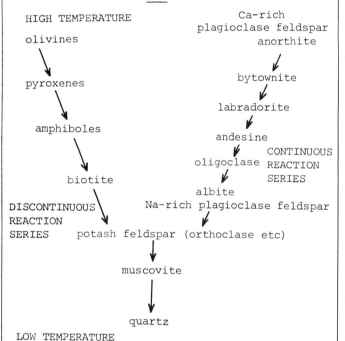

Fig. 3.13 Bowen's Reaction Series.

silicate minerals (Fig. 3.13). One comprises the ferro-magnesian minerals which form a discontinuous reaction series giving rise to the kind of reaction rims illustrated in Fig. 3.7. The other consists of the plagioclase feldspars, which form a continuous reaction series, sometimes seen in zoned crystals. The two series are related to temperatures of crystallisation and unite at their lower temperature ends. The minerals at the high temperature ends of the series (the olivines and calcium-rich plagioclase feldspars) have the highest melting points of their respective series and would therefore be expected to crystallise first as a basic magma cools.

They are also the densest minerals of their series and therefore tend to sink within the melt. These would be followed by the pyroxenes and then the amphiboles with the most acidic minerals such as muscovite and quartz crystallising last.

It is obviously not possible to watch an igneous intrusion crystallising, so one has to use indirect evidence to determine its cooling history. Much of this comes from studies of detailed features such as the zoning of crystals which can be related to the experimental work on magmas already described. There are also a few intrusions which show a systematic distribution of minerals within them which can be related to Bowen's Reaction Series.

The Palisades Sill

The Palisades Sill of New Jersey, USA, provides us with such an example. This sill is over 300 metres thick and it is intruded into Triassic sandstones (Fig. 3.14).

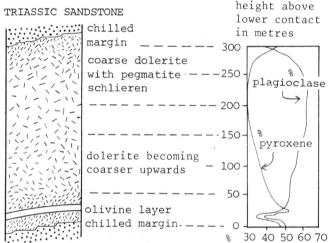

Fig. 3.14 Section through Palisades Sill,
Fig. 3.15 Variation in percentages (by weight) of plagioclase and pyroxenes.

Initially, chilled margins were formed at the top and bottom of the intrusion, each about 16 m thick, and the composition of these rapidly chilled portions of the intrusion is thought to be the same as that of the original magma. This shows it to have been a basic magma with tholeiitic affinities. As the rest of the magma took longer to cool than this chilled margin, it is coarser grained and is thus dolerite rather than the basalt seen in the chilled margin. The dolerite becomes coarser upwards.

As the main part of the intrusion started to cool the silicate minerals crystallised out in the same order as shown in Bowen's Reaction Series. As can be observed from this sequence the olivine and calcium-rich plagioclase feldspars will tend to crystallise first. However, the normal pattern of events whereby these minerals would be expected to react with later liquids has not been followed in the Palisades Sill and a curious distribution of minerals within the intrusion has resulted.

Above the lower chilled margin is a layer 10 m thick which contains 25% olivine - the so-called olivine layer. These olivine crystals have a high specific gravity relative to that of the magma so that after formation they settled out towards the bottom of the intrusion. Some of the crystals may have taken as long as 200-300 years to sink down through the intrusion. Minerals which sink and accumulate at the bottom of intrusions are known as cumulate minerals. Their sinking is one way in which an originally homogeneous magma may begin to crystallise in various fractions, each of which may be significantly different from another. The processes by which such differences are created are known collectively as differentiation.

This trend of differentiation continues throughout the intrusion and although it does not lead to any spectacular layers like the olivine layer, changes can be detected in the composition and proportions of the pyroxenes and plagioclase feldspars. The feldspars become more sodium-rich whilst the pyroxenes increase in iron content in the higher parts of the intrusion. The proportion of the plagioclase feldspars also increases higher in the intrusion (Fig. 3.15). Table 3.2 indicates the contrasts in rock chemistry between different parts of the intrusion.

Because of the early settling out of the more basic components of the intrusion the residual liquid becomes more 'acidic' so that the last part of the intrusion to crystallise, which is about 40 metres from the top, consists of acidic lenses or schlieren containing quartz and orthoclase feldspar. These have crystallised from aqueous solutions which were concentrated in the last part of the intrusion after the crystallisation of the anhydrous phase. They tend to be coarse-grained, which probably reflects the concentration of water and other volatiles as the liquid phase of the magma was gradually displaced upwards by the sinking crystals.

It is worth noting that recent studies have shown that this explanation may be something of an oversimplification. In fact there are 16 metres of normal dolerite between the top of the chilled margin and the olivine layer. This has led to speculation that there may have been more than one pulse of magma intruded, with the olivine settling out at the base of the second pulse.

TABLE 3.2 CHEMICAL COMPOSITIONS OF DOLERITES FROM PALISADES SILL (Weight %)

CONSTITUENT	1	2	3
SiO_2	51.91	48.28	52.32
TiO_2	1.25	0.82	0.97
Al_2O_3	15.31	9.36	16.54
Fe_2O_3	0.98	2.14	1.58
FeO	9.31	11.54	8.66
MnO	0.08	0.12	0.12
MgO	7.52	17.48	5.43
CaO	9.71	7.00	9.68
Na_2O	2.30	1.59	2.32
K_2O	0.79	0.41	1.03
H_2O+	0.93	0.99	0.84
H_2O-	0.15	0.06	0.40
P_2O_5	0.18	0.11	0.06
CO_2			0.16
	100.42	99.90	100.11

1 Average undifferentiated dolerite from chilled margin
2 Olivine dolerite from olivine rich layer
3 Quartz dolerite 150 m above lower contact

The presence of these minor lenses of acid rock in the Palisades Sill is predictable from Bowens Reaction Series but we need to examine an example of a larger intrusion to see if significant quantities of acid rocks can be generated in this manner.

The Skaergaard intrusion

The Skaergaard Intrusion of east Greenland is one of a series of Tertiary intrusions and it is an excellent example of a so-called layered basic intrusion. It has the shape of an inverted cone and, since it has been closely studied by many workers, it is very well documented, although still not perfectly understood.

Around the edges of the intrusion there is a fine grained olivine-gabbro series known as the marginal border group (Fig. 3.16), which is believed to represent the chilled margin

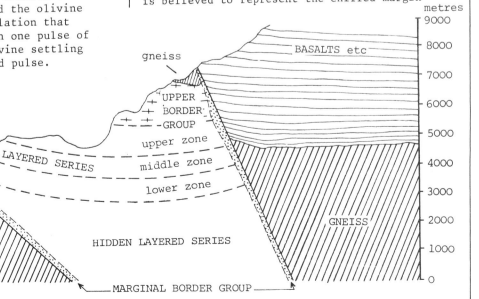

Fig. 3.16 Section through the Skaergaard Intrusion.

of the intrusion, and so, just as with the chilled margin of the Palisades Sill, its composition is thought to be the same as that of the original magma. The marginal border group has suffered contamination due to melting of the surrounding rocks, in particular where it is in contact with gneiss, which has a lower melting point than the Tertiary basalt lavas. However, sufficient samples which are considered to be uncontaminated have been obtained to state with some confidence that the original magma was tholeiitic in character.

The main part of the intrusion is made up of the layered series. As the name suggests this is composed of a succession of layers, each of which is turned up at the edges (Fig. 3.16) and the series represents cooling from the base upwards. By no means all of the layered series is exposed as can be seen from the diagram. Indeed, it has been estimated that over 70% of the intrusion is not exposed and therefore some of the deductions and conclusions reached are, of necessity, rather tenuous.

Within each of the layers, a clear stratification is seen with the dark, heavy minerals, pyroxene and olivine, at the base and the less dense, lighter coloured plagioclase feldspar concentrated near the top of each layer.

This layering is thought to have an origin very much like that of graded bedding in sedimentary rocks. As convection currents swept down through the intrusion they transported crystals with them and deposited them as a 'mush' on the floor of the intrusion. Gravitational settling of the heavy, dark minerals within these cumulate layers would then be responsible for the light and dark bands. Within the layered series, troughs can be seen which have been scoured out by the convection currents.

Less obvious, but of greater significance, is the trend of changing mineral compositions throughout the layered series. Passing upwards, the minerals tend towards the low temperature end members of their respective solid solution series. Thus the ferromagnesian minerals such as olivine and pyroxene become enriched in iron and the plagioclase feldspar becomes richer in sodium. The complete sequence of minerals shown in Bowen's Reaction Series is not seen in Skaergaard as amphiboles and biotite are absent. The intrusion exhibits differentiation to ferrogabbros rather than granites although there are minor quantities of acid rocks in the form of granophyre.

Figure 3.17 summarises the variations in chemical composition of the cumulus minerals. Plagioclase feldspar shows the most complete sequence varying continuously from An_{69} (i.e. 69% anorthite, 31% albite (see Fig. 3.3) at the base of the layered series to An_{33} at the top. Plagioclases of composition An_{77} have been found in the marginal border group, and it is assumed that this represents the composition of the first plagioclases to crystallise. If this is so, plagioclase feldspar of composition An_{77} should be found near

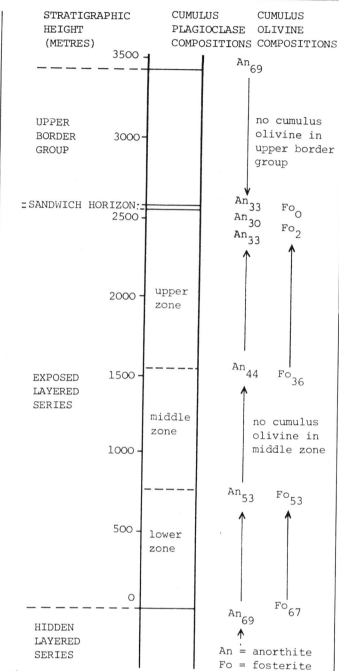

Fig. 3.17 Cumulus mineral composition within the Skaergaard Intrusion.

the base of the hidden part of the layered series. In fact, drilling has confirmed this prediction.

The sequence of chemical variations in the cumulate olivines is not complete since olivine is absent in the middle zone of the layered series. At the base of the lower zone of the exposed layered series the olivine has composition Fo_{67} (i.e. 67% fosterite, 33% fayalite), and it reaches Fo_{53} by the top of the lower zone. When it reappears in the upper zone its composition is Fo_{36} and by the top of the upper zone of the layered series it is Fo_2. When the top of the layered series is reached, the iron content of the rock is great enough for it to be called ferrogabbro.

The third part of the intrusion is the upper border group which, because of erosion, is only seen in part. This is thought to represent cooling from the top downwards since the reverse trend in plagioclase composition compared to that in the layered series has been recorded.

The last part of the intrusion to crystallise was the Sandwich Horizon between the upper border group and the Layered Series. This is the most acidic part of the intrusion and it contains lenses of granophyre with some quartz and orthoclase feldspar, set in quartz-ferrogabbros.

Thus, just as with the Palisades Sill, rocks of acidic composition can be obtained from a basic magma. A comparison of chemical analyses of Skaergaard (Table 3.3) with those given for the Palisades Sill (Table 3.2) shows similar trends although the final products of differentiation are more extreme in Skaergaard. However, in both cases, the amounts of acid rocks produced are very minor. Very slow cooling is necessary for this to happen and it has been calculated that over 12,000 years were needed for the exposed parts of the layered series alone to crystallise. Because of their slow cooling, each of these intrusions has preserved a record of the order of crystallisation of the minerals and has led to a greater understanding of the processes of differentiation working in large plutonic basic intrusions.

TABLE 3.3 CHEMICAL COMPOSITION OF ROCKS FROM SKAERGAARD INTRUSION (Weight %)

	1	2	3	4
SiO_2	47.92	46.37	48.27	58.81
TiO_2	1.40	0.79	2.20	1.26
Al_2O_3	18.87	16.82	8.58	12.02
Fe_2O_3	1.18	1.52	4.06	5.77
FeO	8.65	10.44	22.89	9.38
MnO	0.11	0.09	0.26	0.21
MgO	7.82	9.61	1.21	0.72
CaO	10.46	11.29	7.42	5.03
Na_2O	2.44	2.45	2.65	3.91
K_2O	0.19	0.20	0.34	2.39
H_2O+	0.41	0.29	1.13	0.21
H_2O-	0.10	0.09	0.37	0.19
P_2O_5	0.07	0.06	0.65	0.71
CO_2	0.06			
SrO	0.20			
BaO	0.02			
S	0.27			
	100.17	100.02	100.03	100.61

1 Olivine Gabbro from Chilled margin
2 Olivine Gabbro 500 m above lowest exposed horizon
3 Ferrogabbro 2,500 m above lowest exposed horizon
4 Granophyre from lensoid mass in upper part of intrusion

Granites

Granites and associated rocks such as granodiorite form the largest group of intrusive igneous rocks, comprising over 90% of the total. This contrasts sharply with extrusive rocks where basalt is the most common rock type. Part of the reason for this contrast between the composition of the main intrusive and extrusive rocks has been explained in a previous section on the crystallisation of magma, but part of the answer also lies in the origin of the respective magmas.

Basic magma is produced mainly by partial melting of the ultrabasic mantle. Does the acidic magma which forms granite also come from the mantle or does it have a separate source?

It is possible for small quantities of granitic rock to be formed by differentiation of the primary basic magma as has been illustrated in both the Palisades Sill and the Skaergaard Intrusion. However, in order to produce all the granitic rocks by differentiation, there would have to be at least ten times as much basic magma as acidic magma. This is clearly not the case, as the acidic intrusions are nearly twenty times as abundant as all other intrusions.

As basic magma rises through the continental crust, partial melting may occur and the products become assimilated into the magma,

thus contaminating it. However, it is very unlikely that this process would result in any significant quantities of granites.

If most granites are not formed from primary basic magma, how do they originate? A clue to their origin is provided when the distribution of granites is studied. We find that the vast majority are associated with the continental crust, usually within former orogenic belts. The few small intrusions which are in oceanic areas can be accounted for by differentiation of a basic magma. The chemical composition of the upper continental crust closely approximates to that of granite, and the orogenic areas are regions where temperatures and pressures are abnormally high. Theoretically, therefore, granites could have been formed by the melting of the upper continental crust, but is this possible in practice?

Anhydrous (dry) granite begins to melt at about 950°C-$1,000^\circ$C at the pressures encountered in the crust. As the geothermal gradient, at the time of orogenic activity, could have been as high as 30°C km^{-1}, the temperatures required to produce melting are found at a depth of less than 35 km, which is well within the continental crust in orogenic areas. Hydrous (wet) granitic melt can be produced at even shallower levels in the crust as the temperature required is only about 650°C, which is reached at a depth of 20-25 km. (See Fig. 3.6).

Although we can see that it is perfectly possible for granites to be produced by the heating up and eventual melting of continental crust, the idea has given rise to a long debate between geologists. On one hand there are the 'transformationists' who saw that granite could indeed be produced by trans-formation of continental crust. To the 'transformationists', granite is the end product of metamorphism, produced first as a 'mixed' rock, part gneiss, part granite, known as <u>migmatite</u> which itself undergoes complete transformation at a later stage, becoming granite. On the other hand there are the 'magmatists' who point to the many examples of granites that clearly have formed from an intrusion of magma, and are thus truly igneous. There are points to be made on both sides of the argument and the follow-ing examples illustrate features which suggest a migmatitic origin in some cases and an igneous, or magmatic, origin in others. There then remains the question of whether the genesis of these two types of granite can be connected in some way.

Migmatites and migmatitic granites

As has already been pointed out, it is very common to find granites in orogenic belts associated with areas of high grade regional metamorphism. The sequence of metamorphic rocks - slate → phyllite → schist → gneiss produced from shales or mudstones with increasing grade of metamorphism is well known. Not quite so well known, perhaps, are the properties of the group of rocks known as <u>migmatites</u>.

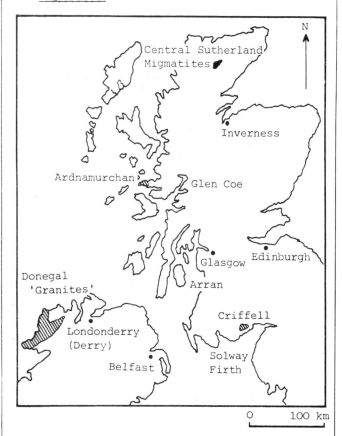

Fig. 3.18 Location of Central Sutherland migmatites and Donegal granites.

These migmatites are most often found associated with schists and gneisses and they are in fact 'mixed' rocks where granitic material has come to be mingled with a host rock which originated by high grade regional metamorphism. The granitic material seems to have migrated through the host, either as magma or aqueous fluids of granitic composi-tion, although it is possible that it took the form of a diffusion of ions migrating through the pore fluids. These are rather loosely referred to as 'emanations'. It is often claimed that these emanations have their origin outside the host, perhaps rising from deeper levels in the crust, although other authors believe that in some cases the granitic material may have originated as a segregation from the host itself.

It is quite easy to see that there are two distinct components in hand specimens, even though metasomatic reactions between them will have caused some modifications in composition. The original composition of the host will have an important part to play in the composition of the final product, but with increasing migmatisation a rock is pro-duced which approaches granite in composition.

Therefore, it has been quite convincingly argued that migmatites provide a link between the true metamorphic rocks and granites and that the complete metamorphic series should read: slate → phyllite → schist → gneiss → migmatite → granite. In Britain the main areas where migmatites are found are in the High-lands of Scotland and in Ireland (Fig. 3.18). Here, during the Caledonian Orogeny, a broad zone of regional metamorphism was formed and migmatites can be seen in such places as the Central Sutherland Complex in Scotland. Here, as in many other examples there is a gradation from an area where the host rock, in this case Moine Schists, is traversed by a few veins of granite, through zones where the granitic component becomes more conspicuous to local occurrences of almost pure granite conforming to the strike of the country rock (Fig. 3.19).

The Main Donegal granite of Ireland shows evidence that it was emplaced by a lateral wedging process yet it has included in it rafts of the original country rock which are not only aligned parallel to the regional trend but also preserve the original or 'ghost' stratigraphy of the area across the granite outcrop (Fig. 3.20). Although, therefore, this is not a migmatitic granite in the sense that it has formed 'in situ', clearly the magma has not moved far from where it originated and it does not show the homo-geneous nature which might be expected of a thoroughly magmatic granite.

Similarly, the older granodiorite, north west of the Main Donegal granite, contains inclusions of Dalradian migmatites which have retained, to a great extent, their original position. The granodiorite appears to pene-trate the country rock but sufficient dis-placement of the xenolithic blocks has been noted for a magmatic origin to be postulated for the granodiorite.

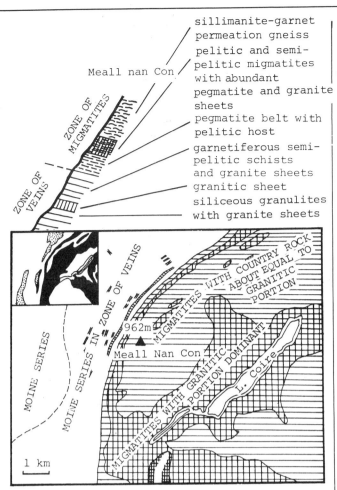

sillimanite-garnet
permeation gneiss
pelitic and semi-
pelitic migmatites
with abundant
pegmatite and granite
sheets
pegmatite belt with
pelitic host
garnetiferous semi-
pelitic schists
and granite sheets
granitic sheet
siliceous granulites
with granite sheets

Fig. 3.19 Regional migmatisation, Central Sutherland. The main map shows the zones of migmatisation; the inset map shows the distribution of original lithological types, black pelites, dotted hornblende-gneiss, unornamented psammites. The section of Meall nan Con (Ben Klibreck) above shows zones of migmatites over zones of veins. The height of the section is about 800 metres and the vertical and horizontal scales are the same.

In contrast, the remaining two granite out-crops in the area are very clearly magmatic, although of different form. The Rosses granites form a ring complex and the Ardara granite is a diapiric intrusion. These different types of intrusions are described later.

An example of a magmatic granite

Many examples of granites which are clearly of magmatic origin could have been chosen from the British Isles; indeed two have already been mentioned in the previous section. In general, they tend to be smaller, more homogeneous bodies than the granites that are believed to have been generated 'in situ' by the processes of migmatisation.

Although not strictly a granite, because of its lower silica content, the main part of the Criffell granodiorite of the Southern Uplands of Scotland provides an excellent example of an intrusion. It is one of a series of Caledonian granite outcrops in the area and it has been forcefully intruded into folded Silurian shales and greywackes. There are three parts to this intrusion, for the main granodiorite grades into a porphyritic

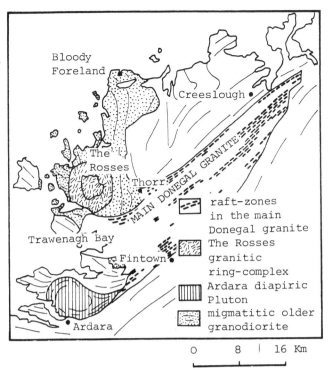

Fig. 3.20 Diagrammatic map of the Donegal Granite.

granodiorite with feldspar phenocrysts in its central portion. The third part is older than the main complex and is seen at its south western end. There, finer grained granodio-rites are associated with some quartz-diorite rocks (Fig. 3.21).

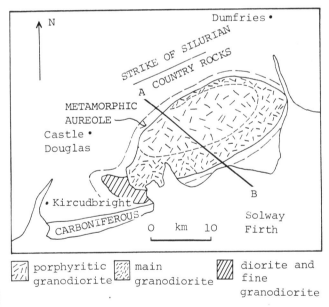

porphyritic granodiorite
main granodiorite
diorite and fine granodiorite

Fig. 3.21 The Criffel granodiorite complex.

The Criffell granodiorite shows a large number of features which could only have originated by the intrusion of hot magma into colder country rocks. Most significantly, in con-trast to the migmatitic granites previously described the contacts with the country rocks are sharp and they dip steeply outwards. This has been confirmed by geophysical surveys across the intrusion. Figure 3.22 illustrates the interpretation by one of the authors after extensive geophysical work in the area. (See the Unit *Geophysics* for further details of the technique.)

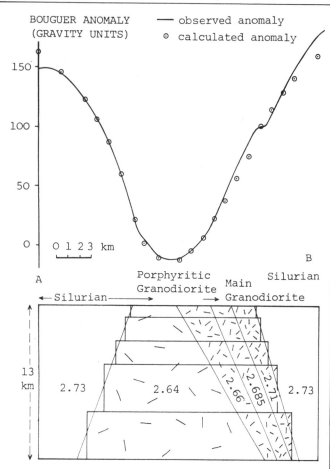

BOUGUER ANOMALY
(GRAVITY UNITS)

— observed anomaly
○ calculated anomaly

Fig. 3.22 Section across a three dimensional model of the Criffel granodiorite showing a good fit between the observed and calculated gravity anomalies. 1 gravity unit = 10⁻⁶ m s⁻². 2.73 etc = rock density in tonnes m⁻³

There is also no gradual transition from high grade regional metamorphic rocks or migmatites, since the country rocks have only previously been metamorphosed to slate grade. The intrusion has, however, formed a clear metamorphic aureole within the country rocks which have been hornfelsed and in some places mineralised. In the outer parts of the intrusion xenoliths are common and veins of the granodiorite invade the country rock in places. A flow foliation has been recorded in the outer parts of the intrusion which indicates that the origin must have been from a mobile magma. Although the outcrop of the granite is slightly elongated in the direction of the strike of the country rocks it is completely discordant to it at its western end which again contrasts with a migmatitic granite.

All these features make it clear that this particular intrusion originated from a hot, mobile magma. It is thought that the mass intruded to a relatively high level in the earth's crust, perhaps rising to within 1,000 m of the earth's surface before finally solidifying.

The granite series

The examples illustrate that some granites result from crustal melting - the so-called 'metamorphic' or migmatitic granites - whilst others have a definite magmatic source - the so-called 'igneous' granites. How can these two greatly contrasting origins be resolved, if indeed they can be resolved at all?

In his book *The Granite Controversy* (1957) H H Read summarised the arguments and proposed what he called the 'Granite Series' which provided a link between the two types of granite. In this series, the metamorphic granites represent deeper levels in the earth's crust where partial melting has taken place, a process known as anatexis. The melt has then consolidated 'in situ'. Sometimes, this melt becomes mobile and, because it has a relatively low density compared to the rocks around it, it will then move upwards to intrude into the higher levels of the crust where it will crystallise to form a magmatic granite. Figure 3.23 summarises the various stages and links between them in the Granite Series.

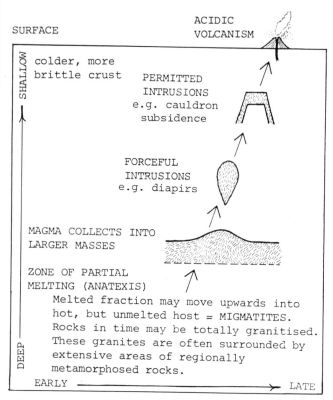

Fig. 3.23 The Granite Series.

One question needs consideration. If most granites can be referred back to crustal melting, how is it that they show little variation in composition when the crust is very variable? Experiments on mixtures of quartz, orthoclase and plagioclase have shown that they begin to melt at varying temperatures, depending upon the relative proportions of the minerals in the mixture. The mixture which remains liquid to the lowest temperatures contains approximately equal proportions of the three components. Analyses of a great number of natural granites, which were poor in ferromagnesian minerals, showed that they contained quartz, orthoclase and plagioclase in much the same proportions as in the artificial mixture. This is because the natural granites will contain the most easily melted fraction, i.e. melted at the lowest temperatures, of the original crustal rocks and so variations in the composition of the crustal rocks will be evened out. Figure 3.24 shows how well the low temperature mixtures match the maximum for the granites.

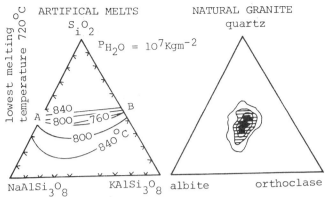

Fig. 3.24 The system SiO_2, $NaAlSi_3O_8$, $KAlSi_3O_8$ and its application to natural granites. (a) The temperatures at which crystallisation begins in a quartz-orthoclase-albite liquid, in the presence of water, at a water-vapour pressure of $10^7 Kgm^{-2}$. Above AB, quartz appears first, below AB feldspar first. The lowest temperatures of crystallisation are shown by mixtures having a composition in the central part of AB. (b) Composition of 571 natural granites containing less than 20% of ferromagnesian minerals, expressed in terms of quartz, orthoclase and albite. Note that the maximum falls in the position of the low-temperature mixtures.

Granites are also found outside orogenic areas where the thinner crust and the lower geothermal gradient means that it is not possible for them to have been produced by crustal melting. For these granites we have to revert to postulating an origin in the mantle. The peridotite mantle must have suffered extreme differentiation to produce a granitic melt. This is theoretically possible, although very large quantities of material would be required, the ratio of original peridotite to granitic melt produced being approximately 100 to 1. Evidence for this hypothesis has recently been found. In crustal granites, and those thought to originate from the mantle, the ratios of two isotopes of strontium have been found to be different, reflecting their different origins. This new work on the strontium ratios has led to increased speculation about the origin of granites. The migmatitic granites are obviously formed in the crust as they are derived from partial melting of crustal rocks. However, magmatic granites could either be derived from partially melted crustal material which becomes mobile and moves upwards as outlined in the Granite Series above or they could alternatively have been derived from the mantle.

The space problem

Several very large granitic masses are known from different continental areas. These are called batholiths and may be hundreds of kilometres in length (Fig. 3.25). Smaller offshoots from the main batholith can occur and these are known as stocks or bosses.

When granites can be inferred to have been formed 'in situ' by an intensification of metamorphic processes leading to crustal melting, there is no difficulty in accounting

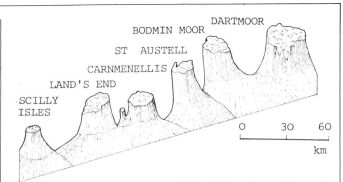

Fig. 3.25 The granite batholith of south west England. The main batholith has a number of smaller offshoots known as stocks or bosses which outcrop at the surface.

for their vast size. The original rocks have merely been altered or transformed. However, when it can be demonstrated that large masses have been formed by the intrusion of magma which came from somewhere else, a mechanism or mechanisms have to be found which can account for the displacement of the existing country rocks. These mechanisms fall into two broad groups known as <u>forceful</u> and <u>permitted</u> intrusions. Forceful intrusions make room for themselves by shouldering aside the country rocks whilst permitted intrusions are formed more passively by magma rising to occupy spaces left by detached masses of country rock. The same principles apply to large intrusions of any composition.

Forceful intrusions

<u>Diapiric intrusions</u> The density of the hot magma tends to be less than the surrounding rocks and so this magma can flow slowly upwards pushing the existing rocks out of the way. In the most simple of cases the rocks are merely bent around the granite forming a dome (Fig. 3.26a) but further upwards motion of the magma will rupture the country rocks and form a <u>diapir</u> (Fig. 3.26b) such as in the Ardara granite of Donegal.

The mechanism of intrusion can usually be inferred from associated structures in the country rocks such as folds formed by the upward and outward pressure of the magma or simply by the dip of the country rock which tends to incline outwards from the granite margin (Fig. 3.27).

These types of intrusion will tend to form 'blisters' on the surface which will make them the target for erosion and so the over-lying cover of country rocks can be quickly removed.

Other types of intrusions which have force-fully pushed aside the country rocks include the large, concordant, sheetlike bodies of laccoliths and lopoliths. Laccoliths cause an arching of the country rocks, much like domes, but less strongly arched. They are more common with the acidic rocks than the basic rocks since the more viscous nature of acidic magmas restricts their flow along the bedding planes and so a thickening occurs around the feeder dyke (Fig. 3.26c).

Lopoliths are mainly associated with basic magmas and are large, saucer-shaped

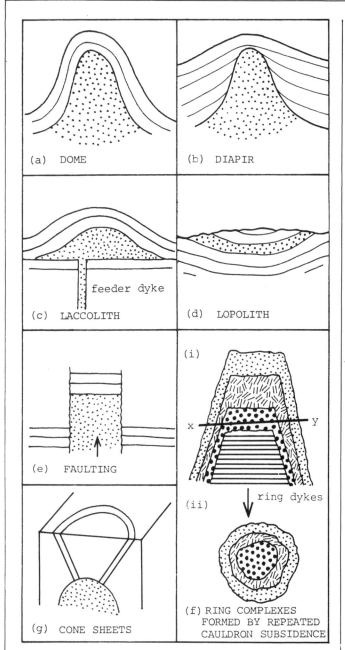

Fig. 3.26 Simplified diagrams to illustrate the various types of intrusions.

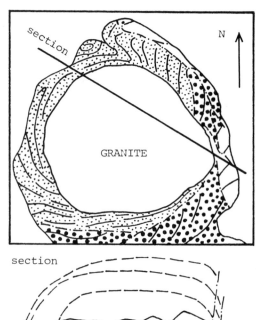

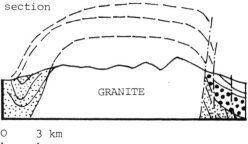

Fig. 3.27 Structures produced in the envelope rocks by the intrusion of the Arran granite dome.

good examples can be seen in Scotland. In fact the process can be multiple with the cauldron sinking repeatedly. At Glen Coe the older Moor of Rannoch granite has been intruded by a second, the Ben Cruachan granite. In the ring complexes of Ardnamurchan in Scotland and the Rosses Complex of Ireland there have been three phases of emplacement by this mechanism (Fig. 3.28).

intrusions. Most appear to have had complex cooling histories and are not, as was first supposed, merely a result of the crust sagging under the excess volume of magma. A very large, well-known example of a lopolith is the Bushveld Complex of South Africa (Fig. 3.26d).

Faulting. Where there are parallel faults, the intrusion makes room for itself by pushing up the country rocks between the faults (Fig. 3.26e).

Permitted intrusions

Cauldron subsidence   There is another more elaborate method by which magmas can be emplaced by faulting. A ring fracture forms, but instead of the central block being pushed up it subsides into the magma and the magma squeezes up the sides of the block, acting as a lubricant helping the block to sink. The magma then occupies the cauldron so formed (Fig. 3.26f). These types of permitted intrusions are relatively common and

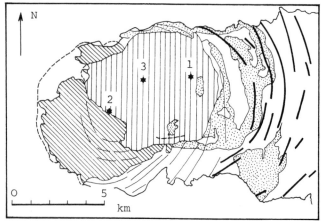

Tertiary basalts
Trias and Jurassic sediments
Moine schists

cone sheets centre 1

cone sheets centre 2

cone sheets centre 3

* positions of igneous centres 1,2,3

centre 1
centre 2
centre 3

Fig. 3.28 Ring complexes of Ardnamurchan. The positions of igneous centres have migrated.

Stoping happens on a smaller scale and enables the magma to digest its way upwards. Lumps of wall and roof rock fall into the magma and they become homogenised with it permitting the magma to move upwards into the space that is left. Evidence of this process can remain as the magma may have a large number of xenoliths in it, where blocks of country rock have not been completely assimilated into the magma.

Minor intrusions

The two most usual types of minor intrusions are sills and dykes. Sills are concordant intrusions, that is, they are sheet-like structures which follow the bedding or other planes of structure. A sill may cross from one plane to another, in which case it is said to be transgressive. There is great variation in the size of sills, the Palisades Sill at over 300 metres thick being a particularly large example. One of the best-known British examples is the Great Whin Sill of Northern England. This averages approximately 30 metres in thickness and it underlies a vast area from the Farne Islands in the north to the Tees Valley in the south. Perhaps it is most well known where its outcrop has formed a north-facing scarp slope along which Hadrian's Wall has been built. Many sills do not reach this size and are found as thin intrusions of no great lateral extent (Fig. 3.29).

Fig. 3.29 *Megaligger Rocks, South Cornwall. Acid sills in shales.*

Dykes are discordant structures where a sheet of magma has cut vertically or nearly vertically across the bedding planes. Dykes can vary greatly in thickness, from a few centimetres, to hundreds of metres across and they can either be of restricted length or run for considerable distances. For instance, the Cleveland Dyke can be traced intermittently all the way from Arran, across northern England to the North Yorkshire Moors. In the Tertiary Igneous Province of north-west Britain, numerous dykes can be seen running parallel to one another in dyke swarms.

A number of minor intrusions are often associated with the major intrusions described above. Commonly, the stress of the magma intrusion will cause fractures in the country rocks and the minor intrusions fill these cracks. A radial pattern of dyke swarms may occur or there may be a concentric pattern known as cone sheets. The cone sheets dip inwards towards the major intrusions with which they are associated (Fig. 3.26g). Figure 2.2. is a photograph showing a cone sheet in the Cuillin Hills on Skye. A radial dyke swarm has originated on Tristan da Cunha (Fig. 1.4) due to stresses built up as magma collected beneath the volcano.

One problem often confronting geologists when they see a layer of igneous rock lying conformably in a succession of sedimentary strata, is to tell whether it is a sill or a lava flow. Perhaps the easiest way to decide is to observe the baked area formed. A sill will have a baked area above and below it but with a lava flow one can only develop in the underlying rocks as there was nothing above it at the time of the extrusion.

Further indications are given by the nature of the upper surface. With a lava flow, this may be uneven and have a soil developed upon it as it could have been exposed to subaerial processes of weathering. On the other hand, a sill could not show this, as it is by definition intrusive. It could, however, contain xenolithic inclusions of the overlying strata whereas it would be impossible for a lava flow to have these.

4
The Global Scene

The Origin of Basic Magma

As we have seen in the previous section, granitic rocks are usually formed from material which is generated within the continental crust, but this is not the case with basic magma. Basalts are frequently found erupting in the ocean basins, far removed from continental crust, and basic magma also tends to be much hotter than acid magma. It usually erupts at temperatures of over $1,000^{\circ}$C whilst it is rare for an acidic lava to exceed 800°C. These higher temperatures are very infrequently reached within the crust. Basic magmas are therefore considered to be derived from the mantle.

For obvious reasons, our knowledge of the mantle is somewhat limited. However, it is possible to obtain information about it in a number of ways. First, xenoliths of mantle material are brought up in kimberlite pipes. These pipes are roughly cylindrical tubes which cut through the continental crust in various parts of the world, notably in South Africa. They are formed of ultrabasic rocks and, as well as the mantle xenoliths, some contain diamonds which show that the pipes must have originated at least 200 km down in the mantle where there is sufficient pressure for diamonds to form.

Further information about the mantle is gathered from the limited number of localities where rocks from it are actually exposed. The closing of the African plate relative to the European has caused the crust in between to be squeezed and buckled. This has become so intense in places that a small part of the upper portion of the mantle has actually been thrust up, folded and exposed at the surface in the Troodos Mountains of Cyprus.

Other evidence concerning the properties and composition of the mantle is obtained indirectly by a variety of geophysical methods such as its response to seismic waves. This reveals a lot about its density and its elastic properties. From the various lines of evidence, it is clear that the upper mantle is a solid with an ultrabasic composition. Originally, it was thought that the upper mantle had a fairly homogeneous composition, but as our knowledge has grown it is seen to be increasingly complex. Nevertheless, fundamentally it is composed of peridotite.

It is from this peridotite that the primary basic magma is thought to originate by partial melting where the mantle temperature becomes locally raised. Even though the upper mantle's composition is variable, the primary basic magma seems to have a reasonably uniform composition. This is related to the way in which partial melting takes place. Once partial melting has started, it proceeds without delay and is able to mix after only a little melting has taken place which helps to produce a homogeneous magma. This melt has a surprisingly low viscosity. Under a pressure of 25×10^{8}Pa, which is that encountered at the depths where the magma originates, the viscosity is only 25 poises. For comparison, glycerine at atmospheric pressure has a viscosity of 100 poises. This melt is therefore very mobile.

The depth at which this partial melting takes place is variable, but a zone which lies at a depth of around 70-150 km, known as the asthenosphere, seems one likely source. This zone has been recognised from seismic studies because the velocity of shear waves reaches a minimum as they pass through it. This suggests that it is a weak zone in which, because of abnormally high temperatures, rocks are close to their melting points. It is even possible that it contains small amounts of liquid. Above the asthenosphere, the upper mantle and crust are more rigid and are known as the lithosphere. The mechanically weak asthenosphere is probably the area in which the lithospheric plates detach from the lower mantle.

This primary basic magma can become modified on its journey to the surface, leading to a variety of different volcanic rocks. There is a broad regional pattern to this variation which will be considered in the next section.

Petrographic Provinces

When the pattern of modern igneous activity is examined, it is very easy to see that there is a regular distribution, the volcanoes usually being associated with active plate margins (Fig. 4.1).

Our earlier studies of Mount St Helens and Tristan da Cunha showed that the type of igneous activity and the nature of its products varies in different parts of the world. However, the pattern of activity is certainly not random and a number of petrographic provinces (also known as magmatic or volcanic provinces) have become recognised. A petrographic province is a broad region where volcanic rocks occur which are genetically related and which belong to a similar period of activity. Each of these provinces

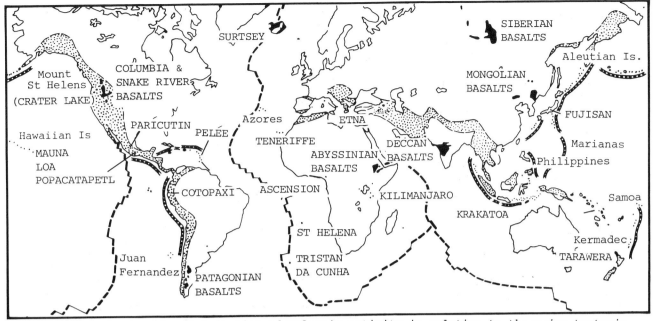

Fig. 4.1 Map showing the distribution of volcanic activity in relation to the major tectonic features of the earth. Positions of the crests of oceanic ridges are indicated by a heavy line, broken to show displacements due to transverse faults. The relationship between oceanic trenches (black and white ornament) and the belts of Tertiary to Recent folding of the continents and island arcs (stippled) are indicated. Dots indicate the positions of only a selection of Recent or currently active volcanoes. The main areas of Tertiary or young flood basalts are shown in black.

will have a limited 'life'. For example the distribution of igneous activity in the Mesozoic would show a very different pattern from today. Figure 4.2 is a schematic diagram showing how magma is generated today within the general framework of plate tectonics, the differences in generation leading to the modern petrographic provinces.

Active ocean ridges

The first province of modern igneous activity we shall consider is the ocean ridge system. This world-wide system of ridges is the site of numerous volcanoes, both submarine and those which reach above sea level and form volcanic islands such as Tristan da Cunha. The ocean ridges are the constructive plate margins. Here, rising convection currents in the mantle make them areas of high heat flow. Partial melting is therefore accomplished at relatively shallow depths. The mid ocean

ridge basalts are tholeiitic where the higher silica content is a reflection of partial melting at shallow depths. As the majority of these lavas are erupted under water, they typically show a pillow structure.

North Atlantic Tertiary Province

In the North Atlantic, the ridge system is still very active and visible above sea-level in places like Iceland. However, in north-west Britain and Greenland there is substantial evidence that these areas were also active when the Atlantic was first opening up in the Tertiary and when, as a consequence, they both lay close to the ridge system. Because of this link between Greenland, Iceland, the Faeroe Islands and North West Britain, they are usually collectively known as the North Atlantic (Brito-Arctic or Thulean) Tertiary Igneous Province.

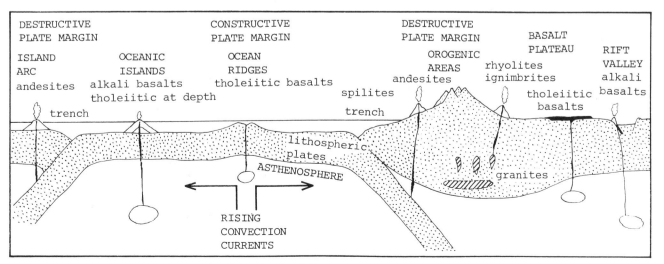

Fig. 4.2. Magma generation and plate tectonics.

Oceanic islands

There are other volcanic islands in the ocean basins which are not connected with the ridge systems, notably the Hawaiian Islands. Here, activity seems to be associated with a 'hot-spot' in the mantle. When the ages of the rocks forming these islands are examined, it can be seen that the dates of activity decrease from the north-west to the south-east, which could be related to the movement of the Pacific Plate to the north-west over such a 'hot spot' in the mantle (Fig. 4.3).

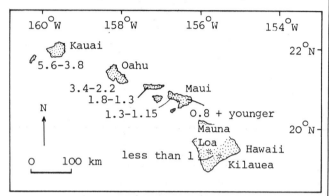

Fig. 4.3. Potassium-argon ages of the Hawaiian Islands in millions of years.

Recent activity, forming lavas in the upper part of the islands, has produced alkali-basalts whilst the vast bulk of the shield volcanoes is tholeiitic. The reasons for the differences in the lava types stem from changes in the magma reservoir. The original tholeiitic lavas may have been produced in areas of higher heat flow, like those found close to the ridge crests but, with time and increasing distance from the ridges, areas of lower heat flow are entered and the lavas would have to be erupted from greater depths resulting in the alkali basalts.

Although the actual mechanisms of generation are still unclear, much detailed seismic work has been completed on Hawaii which gives us information about the rise of magma to the surface. These studies have revealed magma collecting at a depth of 60 km below the surface and over a period of months its slow rise to the surface has been followed. It collected at 2-3 km below the surface just before eruption, causing an expansion of the summit, but after eruption rapid deflation occurred.

Spilites

One other group of igneous rocks associated with the oceanic areas needs to be considered. These are the distinctive basalts known as spilites. Near the continental margins where the basaltic pillow lavas of the oceanic crust are associated with thick accumulations of sediment, the original composition of the basalt has become altered. They seem to have been affected by the introduction of sodium, possibly derived from the sea water trapped within the sediments. The main feature of the spilites is that the basic calcium-rich plagioclase feldspar has been replaced by the sodium-rich variety of plagioclase feldspar,

albite. Chlorite and actinolite also replace some of the original pyroxenes.

Spilites are often found exposed closely associated with ultrabasic rocks. This assemblage is known as an Ophiolite suite. The ultrabasic rocks may represent the top part of the mantle overlain by basaltic oceanic crust, the spilites. This means that when ophiolites are found as in the Troodos Mountains of Cyprus, we may be getting an insight into mantle structure.

Destructive plate margins

At destructive plate margins another distinctive suite of volcanic rocks arises. Typically, andesitic lava is erupted at these margins as a plate is forced down along the subduction zone into the mantle, with its line of descent marked by a trail of earthquakes known as the Benioff Zone. Gradually, the plate will be warmed up as it is forced deeper into the mantle and enormous amounts of heat will also be generated by friction. This will cause partial melting of the top section of the lithospheric plate, the basaltic oceanic crust. As we discovered in a previous section of this Unit, a consideration of Bowen's Series shows that the silicate minerals melt in a definite order. Partial melting can therefore lead to a different product from that formed by complete melting. In this case, andesite is produced by the partial melting of basalt.

The destructive plate margins which surround the Pacific Ocean give rise to a distinctive area of andesitic vulcanicity, the so-called Circum-Pacific Belt or Pacific Ring of Fire. In fact, a boundary between the dominantly basaltic lavas of the ocean basins and the andesitic vulcanicity surrounding the ocean known as the Andesite Line is recognised. The differences between the two petrographic provinces are very significant as they reflect contrasting genesis of the magma. Also, the andesitic vulcanicity tends to be more explosive than that associated with the basaltic lavas.

Andesites erupt on the surface in two situations. Where the plate is subducted beneath oceanic crust, a chain of volcanic islands, known as an island arc, is formed. These curved chains of islands are seen in such places as the Caribbean and Aleutian Islands. Andesitic volcanoes also erupt on the continents and the classic shaped cones typical of this type of vulcanicity are seen in the Rockies and in the Andes where again a plate is being subducted, but this time beneath a continental rather than an oceanic plate. Mount St Helens, referred to in the introductory section, is a volcano of this andesitic type. As a large number of andesites are found in association with continental crust, it was thought the andesites may have originated by contamination of the primary basic magma as it rose through the more acidic crust. Whilst this idea has been largely superseded by plate tectonic theory, it is probable that the contamination process is still responsible for some of the andesites and rhyolites encountered in orogenic areas.

Although andesites are the dominant lava erupted in the petrographic province, basalts are also seen where partial melting has proceeded to a greater extent. Granite batholiths and their associated volcanic products such as rhyolite and ignimbrite are also associated with orogenic areas. However, their probable origins by the melting of contenental crust has already been discussed in some detail in a previous section of this Unit.

Rift valley vulcanicity

By no means all volcanic activity is confined to the ocean basins and margins and there are many extensive areas of volcanic activity in continental regions. The rift valley system of East Africa is a modern petrographic province in which the lavas are distinctively alkaline. The lavas include trachytes and alkaline basalts which contain a greater proportion of silica-deficient minerals, such as nepheline, than normal basalts.

The rift valley may be the site of a future continental split or, as seems more likely, it is an attempt at splitting that failed. The distinctive lavas of this province originate in the upper mantle and make their way to the surface along the fault zones of the rift valley.

Plateau basalts

Also in the continental areas, although not modern petrographic provinces since they have no active counterpart today, are very extensive areas of plateau basalts. Here, vast piles of tholeiitic basalt have built up, erupting from a series of fissures. They are

seen in the Deccan Plateau of India where over half a million km^3 of basalt is found and in the Columbia-Snake plateau of the north-western United States. In Britain, the Antrim plateau is also of this type.

The composition of these basalts is very similar to that erupted at the ocean ridges and so similar conditions of high heat flow and relatively low pressure probably existed. That these basalts erupted from fissures shows that they occurred in areas of crustal tension, as indeed exists at the ocean ridge crests. It has been suggested that these outpourings were started by the initial activities of the spreading ridges. The magma generated spilt out onto the continents before the oceans opened up. The position of many of the lava plateaux on the margins of the present continents tends to support this theory (Fig. 4.4).

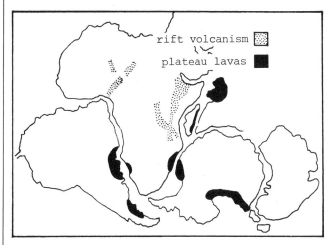

Fig. 4.4 Reconstruction of southern continents showing distribution of regions affected by Mesozoic and younger tholeiitic plateau basalt volcanism and some rift valley features characterised by alkali basalt volcanism.

5

Past Igneous Activity in the British Isles

Although there is no igneous activity in Britain today, there was a great deal in the past. As was dicussed in the previous section, modern igneous activity is largely confined to active plate margins and it therefore seems logical to assume that evidence of igneous activity in Britain in the past must usually be related to times when it was close to a plate margin.

Table 5.1 shows when the main periods of igneous activity are found in the British stratigraphic record. The nature of the activity can then be related to the modern petrographic provinces so that the type of plate margin causing the activity can be deduced.

It is very difficult to reconstruct what plate motions may have been like in the Precambrian and it is likely that, in the early stages of the formation of the crust, the plate tectonic activity was not as we know it today. Since then, however, a fairly clear pattern of Britain's plate tectonic history emerges.

Throughout the Lower Palaeozoic there was closing of the Iapetus Ocean, with northern Britain being on the same side as Greenland and North America, and the rest of Britain on the other side. Following the closing of this Ocean, which is marked by the Caledonian orogeny, Britain became part of a stable plate, continental to begin with but later covered by shelf seas. In the Upper Palaeozoic, the main phase of igneous activity was connected with the Midland Rift Valley of Scotland, although the southern part of England was affected as the ocean to the south of Britain closed causing the Armorican (or Hercynian) orogeny. In the Mesozoic, igneous activity was more restricted as Britain remained part of a stable plate. The activity was connected with the break-up of the northern continents. The first attempt at splitting seems to have been in the North Sea but eventually the con-

TABLE 5.1 AN OUTLINE OF PAST IGNEOUS ACTIVITY IN THE BRITISH ISLES.

PERIOD	LOCALITY	NATURE OF ACTIVITY	PLATE TECTONIC INTERPRETATION
QUATEPNARY AND TERTIARY	W. Scotland and N. Ireland	basalts dykes and plutonic centres	constructive plate margin: opening of Atlantic
CRETACEOUS	nil		
JURASSIC	North Sea	lavas	associated with some faulting: an attempt at a constructive plate margin
TRIASSIC	nil		
PERMIAN	Midland Valley of Scotland S.W. England ———	basalts basalts and major granites	rift valley destructive plate margin ———
CARBONIFEROUS	Midland Valley of Scotland S. Pennines and S.W. England	basalts and trachytes some local volcanic activity	rift valley
DEVONIAN	S.W. England Midland Valley of Scotland Cheviot Hills Highlands and Southern Uplands of Scotland ———	basalts and tuffs andesites and dykes andesites and granite andesites, tuffs, major ——— granites and gabbros	destructive plate margin: final closing of Iapetus Ocean
SILURIAN	Lake District insignificant vulcanicity		
ORDOVICIAN	Lake District North and S. Wales S.E. Ireland Southern Uplands	andesites, tuffs granites andesites, rhyolites, tuffs tuffs, lavas, basalts	destructive plate margin: island arcs on edge of closing Iapetus Ocean
CAMBRIAN	nil		
PRECAMBRIAN	There are many outcrops of igneous rocks in the Precambrian, e.g. lavas and tuffs in Shropshire, Leicestershire and the Malvern Hills; spilitic basalts and granites in Anglesey; dolerite dykes in N.W. Scotland; peridotite in the Lizard Peninsula. It is difficult to place these into a plate tectonic framework because of their age and often deformed state, but there may have been a destructive plate margin in the late Precambrian in Anglesey.		

structive plate margin formed west of the
British Isles, in a different place from the
original junction formed by the closing of
Iapetus. This constructive plate margin
resulted in the extensive areas of Tertiary
intrusive and extrusive igneous activity of

Western Scotland and Northern Ireland.

Two examples of past igneous activity in
Britain will be examined in more detail to
illustrate the parallels they show with
activity at modern plate boundaries.

Ordovician Vulcanicity

Evidence of Ordovician volcanic activity is
seen principally in North Wales and the
Lake District and to a lesser extent in the
Southern Uplands of Scotland. The overriding
characteristic of this volcanic activity is
found in the andesitic lavas with a high
percentage of pyroclastic rocks, which
immediately suggest formation at a destructive plate margin similar to the circum-
Pacific Belt of today. But is there any
further confirmation of this?

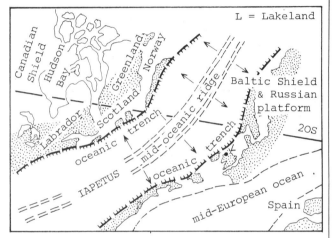

Fig. 5.1 Iapetus Ocean in the Lower Ordovician.

An ocean between England and Wales ('Europe')
and Scotland and North America ('America')
(Fig. 5.1) has long been suspected on
palaeontological grounds. For instance, the
trilobite and brachiopod fauna of England
and Wales differs significantly from that
found in Scotland, especially at the beginning of the Lower Palaeozoic. The Scottish
fauna, on the other hand, has close affinities with that of North America. The sub-
duction of oceanic crust at a destructive
plate margin provides an answer to what
happened to this ocean called Iapetus or the
Proto-Atlantic.

In the Girvan-Ballantrae area of the Southern
Uplands, there are spilites exposed which are
thought to be a relic of this ancient ocean
floor which was thrust up between the closing
continents. Just where the suture between
'Europe' and 'America' lies is still a
matter for some debate but it probably lies
along the Solway Firth, making the Lake
District the leading edge of the European
plate.

Most of the evidence of volcanic activity of
this period is preserved on the European
side of the ancient ocean. In the Lake
District, the first phase of activity is the
Eycott Group. This group contains andesites
and basalts of tholeiitic types which are

very much like lavas found at continental
margins or island arc volcanoes today. The
character of the lavas indicates that they
were erupted through continental crust and
this has been confirmed by a recent geophysical survey of Britain. This shows there
is continental crust to an average depth of
35 km beneath the area. (See the Unit
Geophysics.)

The Eycott Group was followed by a thicker
sequence of volcanic rocks, the Borrowdale
Group. This group includes basaltic and
andesitic lavas with a high proportion of
tuffs. Some of the acidic tuffs are ignim-
britic. Many of the lavas are brecciated,
indicating explosive eruptions, and this
feature is commonly seen in andesite
volcanoes in South America today. However,
the closest modern parallel to the type of
activity seen in the Borrowdale Group is
found in the Cascades of North America. This
is part of the western Cordillera and
includes Mount St Helens. Thus, a similar
geotectonic situation must have existed in
Ordovician times as exists in the East Pacific
today, oceanic crust being subducted beneath
continental crust and the consequent volcanic
rocks rising through the continental crust
(Fig. 5.2).

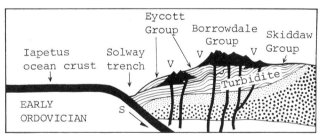

Fig. 5.2 Origin of volcanic rocks at a
destructive plate margin in the Ordovician.

Similar intermediate and acidic lavas and
tuffs are found in the Ordovician rocks of
North Wales. There are clear indications, as
indeed there are in the Lake District, that
whilst some of the eruptions were subaqueous,
other volcanoes built themselves up above sea
level as the lavas lack a pillow structure
and the ash seems to have settled out subaerially. Marine microfossils have been found
in mudstones interbedded with the first tuffs
which confirms that the volcanoes built themselves up through the sea rather than through
lakes.

Putting all the evidence together, the
Ordovician volcanic rocks of North Wales and
the Lake District are clearly an ancient
series of arcs underlain by continental
crust and formed along the continental margin
of ancient Europe.

Tertiary Vulcanicity

A contrast to the Ordovician vulcanicity is found in the Tertiary. Here the igneous activity was not related to a destructive plate margin, but rather to a constructive plate margin.

This province covers a vast area from Greenland to Iceland to Britain to Spitzbergen. Part of the province is, in fact, still active, namely central Iceland. Many of the rocks of this Tertiary Igneous Province have been referred to in previous sections. The Skaergaard Intrusion of East Greenland, the Antrim Plateau of Northern Ireland and the Ardnamurchan Ring Complex of Western Scotland are all of Tertiary age (Fig. 5.3).

The geotectonic affinities of this province are perhaps easier to appreciate than for the Ordovician as the constructive plate margin is still active in Iceland. However, a more detailed look at the Tertiary igneous rocks of Britain shows quite a complex history of intrusions and extrusions related to the opening of the Atlantic.

The initial activity in the area was the outpouring of massive amounts of basaltic lava which are seen in Antrim, and on Mull and Skye. On Mull over 2,000 m of lava are still preserved and many more metres must have been removed by both contemporaneous and recent erosion. These lava piles are made up of many individual flows varying in thickness up to 50 metres but generally 12-15 m thick. The lava flows may show columnar jointing where there has been a slow, regular cooling. The classic examples are seen at Giants' Causeway in Antrim and Fingal's Cave on Staffa, although there are many other examples.

Many of these lavas were erupted subaerially and the small amount of pyroclastic rocks shows that there was little explosive activity associated with the eruptions. This placid outpouring of basaltic lavas, many with tholeiitic affinities, is consistent with activity at constructive plate margins. The difference here is that the magma rose through continental crust to reach the surface as the two continents had not yet split apart.

In addition to the extrusive activity, there are many intrusions in the province which postdate the lavas. These intrusions are referred to as the central complexes.

Many of the central complexes comprise ring intrusions (see section 3) which are the lowest levels of the volcanoes. They were therefore intruded to a high crustal level, although many are coarse-grained. The rock types in these intrusions are both acid and basic. Ardnamurchan (Fig. 3.28) provides a fine example of a basic ring intrusion but many other igneous complexes are seen, for example on Skye, Rhum, Mull, in Northern Ireland and in the granite intrusion of Arran (Fig. 3.27).

The largest basic complex is the Cuillin Hills of Skye where the banded gabbros show a layered structure similar to the Skaergaard Intrusion. Adjacent to the Cuillin Hills lie the Red Hills formed of the acidic rock, granophyre. These are younger than the Cuillin Gabbro and it seems likely that they owe their origin to a melting of continental crustal rocks such as the Torridonian arkoses or the Lewisian Gneiss. The melting of these rocks was caused by the intrusion of large volumes of hot basic magma into the continental crust. Other acidic intrusions in the area would have had a similar genesis.

Many minor intrusions are found within the Tertiary Igneous Province and the origin of cone sheets and radial dykes has been discussed in a previous section.

However, the most revealing minor intrusions, in terms of the geotectonic origins of this province, are north-west to south-east dyke swarms (Fig. 5.3). These lie parallel to the original ridge crest and therefore to the major direction of tension within the crust at the time of the opening of the Atlantic. Their origin is therefore consistent with formation at a constructive plate margin.

The Tertiary igneous activity in Britain is confined to the Eocene, the first part of the Tertiary, and since then Britain has become inactive as the Atlantic opened up and it moved further away from the source of activity. All the igneous features observed are consistent with the initial splitting of two continents. It is interesting to note that when the Atlantic opened up it did so in a different place from the original junction of Iapetus. Scotland consequently became part of Europe rather than North America.

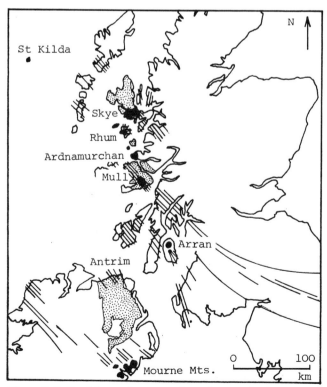

Fig. 5.3 Tertiary Igneous Province of the British Isles. Plutonic centres in black, plateau lavas dotted and dyke swarms marked by lines.

Answers

Crystallisation of plagioclase feldspars (p. 14)

a) Under very slow cooling conditions there is time for equilibrium to be maintained between the melt and the crystals. The first crystals to form are rich in anorthite, but as the liquid cools it becomes progressively richer in albite. It would then be out of equilibrium with the crystals and so a reaction takes place between the melt and the crystals until they are once more in equilibrium. The crystals which are finally preserved in the igneous rocks are thus uniform in composition throughout, but are more albite-rich than they were at the start of the crystallisation.

b) The ions of calcium and sodium are of comparable radius (Ca^{2+} = 0.99Å : Na^+ = 0.97Å), although of different valency. There is therefore enough space in the silicate lattice to allow substitution of one element for the other. The difference in valency is relatively easily compensated for by variation in the proportion of aluminium (Al^{3+}).

c) Zoning is produced when cooling is too rapid for equilibrium to be established between early-formed crystals and the melt. Instead of complete reaction between them, layers of increasingly sodium-rich feldspars accrete around the calcium-rich core.

Crystallisation of acid magma (p.16)

The 'wet' acid magma in the simplified example would rise to about 7 km depth in the crust before crystallisation as an <u>intrusive</u> mass. It would <u>not</u> reach the surface as lava.

Textures of igneous rocks (Table 3.1 and Fig 3.7)

The answers to the 'matching' exercise are as follows:

devitrified obsidian	b
euhedral grains	hI
subhedral grains	hII
anhedral grains	hIII
granophyric texture	e
graphic texture (in a Norwegian graphic granite)	a
zoned crystals	d
'corona' structure	g
ophitic texture	f
porphyritic texture (in a trachyte from the Drakenfels, Germany)	c

Interpretation of textures (Table 3.1)

Graphic texture

The constant ratio of quartz to feldspar and the intergrowth of crystals suggest crystallisation of a melt containing the two components in the eutectic proportions. The shapes are controlled by the atomic structures of the two minerals.

Ophitic texture

The feldspars started to crystallise first around widely separated nuclei in the melt, before they were joined by the pyroxene which crystallised slowly, forming large plates.

Porphyritic texture

The more obvious answer is that the phenocrysts started to crystallise slowly at considerable depth below ground. Then the magma rose to a higher crustal level, or was erupted as lava, when the rest set quickly to produce a finer-grained ground mass. However, in some plutonic rocks, it is possible that the phenocrysts grew during the late stages of the emplacement of the magma, under the influence of enriched volatile components.

Texture of the rock in Fig. 3.12

The ground-mass is fine-grained and dark coloured. The oblong crystals are euhedral phenocrysts of feldspars which appear to be zoned. The round dark blobs are vesicles infilled with a dark green mineral, probably chlorite. The round white blobs (centre) are calcite-filled vesicles.

The rock is an amygdaloidal, porphyritic basalt of Ordovician age from Builth Wells in Wales.

Further Reading

Hatch, F.H., Wells, A.K. and Wells M.K. *Petrology of the Igneous Rocks*. Thomas Murby & Co. 1972. An undergraduate text of long standing, last revised in 1972. Gives thorough systematic study of igneous rocks, followed by a useful section on former igneous activity in the British Isles.

Holmes, A.
Holmes Principles of Physical Geology. Nelson, 1978. The third edition of this well-known text gives dramatic accounts of volcanic activity and also contains a systematic survey of igneous rocks in general.

Institute of Geological Sciences
Volcanoes. H.M.S.O. 1974. A profusely illustrated, colourful guide to vulcanicity, ancient and modern. Intended for the well-informed general reader.

Open University
Internal Processes. S23 Block 4. Open University Press, 1972. The igneous rock section of this workbook links igneous activity to the physical chemistry of magmas and keeps the reader 'on his toes' with questions.

Simpson, B.
Rocks and Minerals. Pergamon Press, 1966. Provides a broad view of rocks of all origins, and also describes the mineralogy of their constituents.

Metamorphism

Contents

Acknowledgements

The authors are most grateful to the following for their help in the preparation of this Unit: Dyson Refractories Ltd. Sheffield for information regarding the manufacture of silica bricks and for the specimens in Fig. 1.1; Dr P.S. Doughty of the Ulster Museum for kindly allowing, free of copyright, the use of the Photograph from the R Welch Collection figured in Fig. 1.2; Dr A.G. Fraser and Dr F. Spode for devoting their time to reading the manuscript and making so many valuable comments on technical matters; Mrs A Kay and Mrs B.G. Ross for their careful typing.

We are grateful to the Institute of Geological Sciences for permission to reproduce Figs 4.1 and 4.6.

The following figures have been based on illustrations from the sources indicated: Figs 3.1, 3.4, 3.9, 3.11, 4.9, 5.4, Harker, A. *Metamorphism*; Figs 3.15, 4.3, 4.4, 5.15, 6.1, Mason, R. *Petrology of the Metamorphic Rocks*; Fig. 4.2, Holmes, A. *Principles of Physical Geology*.

Foreword

Except where otherwise acknowledged, the photographs have been taken by the authors. The scale bar is 5 cm long.

Units of measurement: We have used the derived S.I. unit of pressure, the pascal (Pa) throughout. 1 Pa is equal to 1 newton per square metre. The older unit is the kilobar. 1 kilobar is equal to 10^8 pascals.

Work suggestions have been printed in italic type and it is intended that you work out the answers as you go along.

1

Introduction

Man-made Metamorphism

Most of us owe our homes to metamorphism! Unless you live in a steel and concrete flat, the chances are that your house will have been built of brick and quite possibly roofed with baked clay tiles. Both these building materials were produced by working wet clay into the appropriate shapes and firing them in a kiln. We can regard the transformation from soft clay to hard bricks and tiles as a type of metamorphism brought about by heat, where reactions have taken place which have resulted in permanent physical and chemical changes. If the reactions were readily reversible, the products would be of no value as building materials: imagine the problems in the British climate if bricks reverted to clay as soon as it rained! Instead, they weather almost imperceptibly mainly by mechanical processes, and may take centuries before they are broken down into separate particles.

Evidently a certain minimum temperature must be achieved for such 'metamorphic' changes to take place. In sunnier climates bricks are frequently made by allowing the clay blocks to dry in the sun. The resulting bricks are quite hard, but need protecting from tropical downpours by wide overhanging roofs or long gutters which shoot rainwater well clear of the walls.

In a sense, the drying of the sun-baked brick is akin to some of the processes which take place near the earth's surface when a sediment undergoes lithification to become a sedimentary rock. The kiln-firing is equivalent to one of the ways in which a rock may be metamorphosed.

Although a brick maker knows the precise temperature needed to effect the necessary changes in the clay, in the natural world the boundary between lithification and metamorphism is an ill-defined one.

An ordinary house brick is not a very photogenic subject, but the changes brought about by kiln-firing a silica-brick provide clearer illustrative material (Fig. 1.1a and b). Silica bricks are produced for blast furnace linings and similar refractory uses in the glass industry. They are made from a mixture of siliceous materials such as chert, ganister, silica sand and quartzite, ground down and brought to the right consistency with water and a minimal amount of clay. The mixture is forced into a mould and impressed with the manufacturer's trademark. Figure 1.1a shows a 'green' brick from the mould ready for firing.

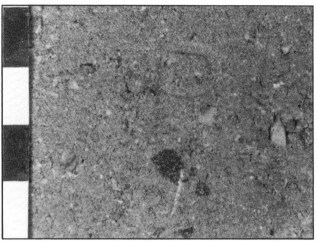

Fig. 1.1a Part of the surface of a 'green' silica brick before firing.

In the traditional process, the green bricks are loaded into a kiln, the temperature of which is slowly raised to 1450°C over a period of about 9 days. This temperature is sustained for a further 4 days and the kiln is then allowed to cool for about 10 days before it can be entered. The firing temperature inside the kiln is far higher than that encountered in natural metamorphism and results in some local partial melting of the constituents. The bulk of the brick, however, remains solid during the firing process. Of course, the time during which the bricks are in the kiln is far shorter than that available in natural metamorphism.

When the bricks emerge they appear like the one in Fig. 1.1b, which has been sawn, ground smooth and photographed at the same scale as Fig. 1.1a. Not only is the brick much harder, stronger and less permeable, but it has also changed colour. Instead of

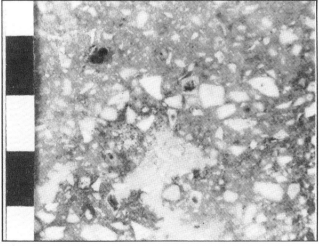

Fig. 1.1b Sawn surface of a silica brick after firing at 1450°C.

the drab green-brown, it now displays variegated colours, ranging from orange to white. Fig. 1.1b looks as though it is out of focus, but this is not so. The blurred outlines of the chert fragments indicate that they have recrystallised to form another type of silica. There has probably also been some migration of ions between the constituents of the brick. Experiments suggest that this is greatly accelerated by the presence of water in the pore spaces of the green brick, acting as a carrier. Under the atmospheric pressure at which the kiln operates, much of the water would be released into the atmosphere as vapour, but this is not necessarily the case under the higher pressures which are found deeper in the earth. The presence of pore water is of the greatest importance in natural metamorphic processes, where there is frequently greater diversity of components than in the brick.

The silica brick is analogous to many natural systems in another way, too. Apart from the loss of some of the water, there has been no change in the bulk chemistry of the brick during its firing: new elements have neither been added to the brick, nor lost from it. Such a change is termed isochemical, and the majority of metamorphic transformations are of this type. Metamorphism in which there is an interchange of material between one body of rock and another is called metasomatism, as when a granite magma invades a limestone.

Metamorphism in the Natural World

Fig. 1.2 Basaltic dyke cutting Chalk and overlying Tertiary lavas. Cave Hill, Belfast. (R. Welch Collection; courtesy of Ulster Museum.)

The manufacture of silica bricks provides us with a kind of controlled experiment in metamorphism. Whereas in the natural world we can observe sedimentation and even some types of igneous activity, it is not possible to see metamorphic processes at work in nature. Yet there is sufficient evidence among rocks now exposed at the earth's surface to show that they were once under much higher pressures or temperatures than they are now and that some degree of recrystallisation has occurred in the solid state since their sedimentary or igneous origins. The first geologist to appreciate the nature of such rocks was Charles Lyell in the 1830s and it was he who coined the name 'metamorphic'. Much of this Unit will be devoted to detailed studies of the metamorphic changes which may be seen in rocks, but we can usefully summarise the main lines of evidence here in three categories.

1. Rocks situated close to undoubted igneous rocks of younger age show marked changes in texture and degree of crystallin-

ity. Superb examples are to be seen where lavas of Tertiary age overlie the Chalk in Northern Ireland, or where dykes of basalt cut through the Chalk (Fig. 1.2). Where the Chalk contacted the once hot magma it was altered to crystalline marble for a distance of several centimetres. This type of metamorphism is known as contact, or thermal metamorphism.

2.   In regions where mountain building has occurred, it has frequently happened that huge masses of rock have been transported for several tens of kilometres by major thrusts. It is only to be expected that some of the rock lying along the thrust plane will have been altered almost beyond recognition. In such environments, lens-shaped fragments of the pre-existing rock are found, embedded in a recrystallised matrix showing a characteristic texture which is not seen in either sedimentary or igneous rocks.   The name dynamic, dislocation or cataclastic metamorphism is applied to this kind of process.

3.   Unlike the last two categories, which represent comparatively localised occurrences, and can be related to particular intrusions or faults, another type of metamorphism seems to have affected whole sequences of rocks occupying wide expanses of country. Growth of new crystals in earlier-formed rocks has quite clearly occurred, frequently demonstrating progressively more complete recrystallisation as the region is crossed. The rocks often indicate a 'lining-up' of flaky minerals which can sometimes be shown to be quite unconnected with original bedding structure. Figure 1.3 shows a slate which cleaves perfectly along the clearly defined plane which lies parallel to the page. The bedding of the original sediment is shown by subtle changes of colour and lies nearly at right angles to the cleavage. In the field, the cleavage direction can often be shown to be related to the intense folding which the

Fig. 1.3 Slate showing cleavage planes almost at right angles to bedding of the original sediment. Charnwood Forest Leicestershire.

rocks underwent and must therefore be due to a process later than the deposition and lithification of the sediment (see p.24). Other, more highly crystalline rocks appear to be related to similar processes. Because this kind of metamorphism affects rocks over wide areas, it is known as regional metamorphism. We shall show later that it results from notable increases in both temperatures and pressures.

Thus, although we cannot apply the principles of uniformitarianism to metamorphic rocks in the same way as we can to most sedimentary and some igneous ones, we still have several powerful lines of evidence on which to base our understanding of metamorphism. The importance of the field relations of metamorphic rocks, their new minerals, textures and structures have been hinted at above. These will be considered in detail when we study each of the main categories of metamorphism. However, we shall discuss first the place of laboratory experiments which aim to reproduce some of the conditions of metamorphism.

2
Laboratory Experiments

One of the earliest geologists to experiment with metamorphic reactions was Sir James Hall who sought in 1805 to imitate the formation of marble. He was aware that when limestone is heated in a limekiln it is converted to quicklime and carbon dioxide gas is given off. The reaction can be expressed as:

$$CaCO_3 \rightleftharpoons CaO + CO_2$$

calcite $\rightleftharpoons$ quicklime + carbon dioxide

Natural occurrences of quicklime, or its hydrated equivalent, are virtually unknown, yet marble outcrops are common wherever igneous rocks have come into contact with limestone or chalk. Perhaps heat is not the only variable!

Hall reasoned that the result might be quite different if the carbon dioxide were not allowed to escape, so he devised a pressure-resistant 'bomb' and filled it with pulverised chalk. The bomb was sealed and heated in a glass-furnace. Sure enough, the chalk was transformed into a crystalline marble. The relatively small increase in pressure had forced the reaction to the left (see equation above) but the CO_2 molecules which were temporarily released had aided the growth of $CaCO_3$ around the existing grain boundaries. The pressure which had built up in the bomb was not high – perhaps only as much as would be produced in nature by a few tens of metres of overlying rock, but it was sufficient to confine the CO_2.

Simple though this experiment was, it paved the way for more modern work, using highly sophisticated and expensive apparatus.

Figure 2.1 is a diagrammatic section through a piece of equipment where water is applied under high pressure to a sample contained in a bomb. This is heated in a furnace to a controlled temperature. Such apparatus enables one to vary not only the pressure and temperature but also the mineral components and fluid content of the sample. It cannot, however, imitate geological time! Because of the nature of the equipment, it is most suited to work with closed systems, but since we have already suggested that the majority of natural transformations appear to have been isochemical, it is a useful parallel.

Most recent experimental work has been directed at determining the pressure/temperature conditions under which particular minerals grow in the solid and are most stable. The results are usually shown in the form of a graph where the stability fields of the minerals are shown. A good example is provided by the effects of pressure and temperature on aluminium silicate, $Al_2 SiO_5$. This exists in three distinct mineral forms, andalusite, kyanite and sillimanite but all three have the same formula. They are referred to as <u>polymorphs</u> of each other. The usually accepted stability fields are shown

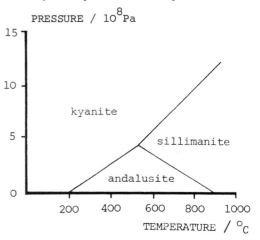

Fig. 2.2 Stability relationships of andalusite, kyanite and sillimanite in the Al_2SiO_5 system (after Vernon, 1976).

in Fig. 2.2. Each mineral is stable within the field where its name occurs, or along the line which divides it from the other minerals in the system. It may exist on the other side of the line, but it will not then be stable.

Unfortunately, different researchers have obtained different results! Figure 2.3 shows a selection of curves, all purporting to be for the same reactions, but demonstrating great diversity. This means that we cannot

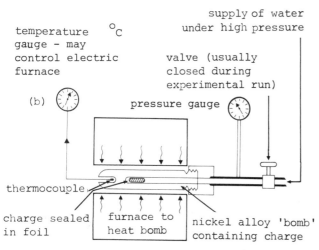

Fig. 2.1 Diagram of a high pressure/temperature apparatus used in laboratory experiments in metamorphic reactions.

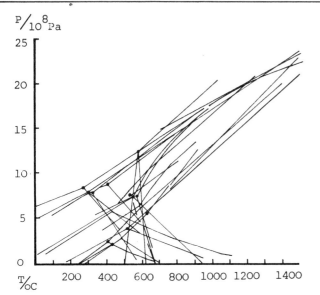

Fig. 2.3 The Al_2SiO_5 reaction curves in positions inferred by various experimenters (from Vernon 1976).

afford to be dogmatic about the precise conditions experienced during metamorphism; nonetheless, the graphs are useful in a semi-quantitative way. Figure 2.2, for example, enables one to predict that andalusite will be the most likely form of Al_2SiO_5 to be found in the relatively high temperature, low pressure environments encountered in the country rock next to an igneous pluton. Kyanite would be expected to form under the higher pressures of regional metamorphism and sillimanite under the most extreme conditions.

The gradients of the curves in Fig. 2.2 indicate that pressure and temperature act together. There are, however, some meta-morphic reactions which are much more depen-dent upon one variable than the other. Such reactions can be used as geological thermom-eters, or geological pressure gauges, once the laboratory work has been carried out.

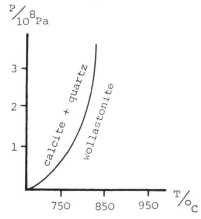

Fig. 2.4 Stability relationships of wollastonite and calcite + quartz (simplified).

Figure 2.4 shows a reaction which can be used to derive a minimum figure for temperatures through which the system must have passed. The equation for the reaction is:

$$CaCO_3 + SiO_2 \rightleftharpoons CaSiO_3 + CO_2$$

calcite + silica $\rightleftharpoons$ wollastonite + carbon dioxide

In this case, the reaction is not so readily reversible as in the marbling of limestone and the CO_2 is driven off.

Up to a pressure of about 2×10^8 Pa (equivalent to the relatively shallow depth in the crust of 7 km) the reaction is both pressure and temperature sensitive but after this point changes in pressure have no appreciable influence. Therefore, the presence of wollastonite in a rock which was formerly composed of calcite and silica indicates that the temperature of the rock must have been raised to at least 800°C.

It is difficult to find an equivalent reaction which is dependent only on pressure changes and not on temperature, but there are some transformations which take place at relatively low temperatures, so long as the pressure is high enough. An example is the reaction whereby the plagioclase feldspar, albite, is converted to jadeite (a pyroxene mineral) and quartz, i.e.

$$Na\,Al\,Si_3\,O_8 \rightleftharpoons Na\,Al\,Si_2\,O_6 + SiO_2$$

albite $\rightleftharpoons$ jadeite + quartz

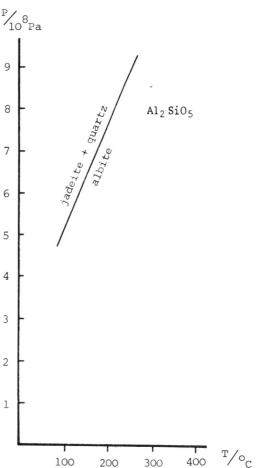

Fig. 2.5 Stability relationships of albite and jadeite + quartz (simplified).

The reaction curve is shown in Fig. 2.5. In nature, there are probably few metamorphic reactions which take place below about 300°C, so continuing the reaction curve up to this temperature indicates that the pressure must be at least 10^9 Pa for jadeite to become stable. This is equivalent to about 35 km depth in the earth.

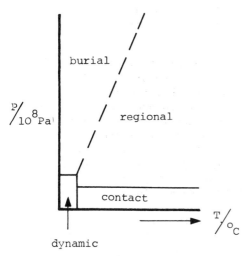

Fig. 2.6 Pressure/temperature fields for the main types of metamorphism.

Many other similar reactions have been imitated experimentally and their behaviour extended from the laboratory work by the application of chemical calculations. This enables us to determine the main pressure/temperature conditions under which metamorphism might have taken place. As shown earlier, the field relations of metamorphic rocks indicate that there are three general environments in which they may be formed: in contact with igneous rocks, along major fault planes and on a regional scale. These broad areas may be loosely related to the pressure/temperature axes as shown in Fig. 2.6. The dividing lines between the areas must be thought of only as general guidelines and not as hard and fast rules. The graph shows an additional category: that of burial metamorphism. This includes the albite/jadeite transition described above. Some geologists prefer to make this distinction between rocks produced by deep burial and those arising from other processes of regional metamorphism. The use of the latter term usually implies that strong lateral stresses were acting, perhaps in addition to considerable depths of burial.

We shall now consider each of the main categories of metamorphism in turn. In each section we shall examine the evidence accruing from changes in mineralogy, texture and structure within the rocks. In many cases, it is also possible to use relict features to determine their original nature.

3
Contact Metamorphism

Figure 1.2 shows a situation where a magma has forced its way through a country rock of Chalk and poured out on top as a lava. Later, a fresh injection of similar magma produced the basaltic dyke cutting both Chalk and lavas.

The effects on the surrounding rocks of such igneous activity depend on several factors. These include:

1. the temperature of the magma when it reaches the country rock
2. the size of the igneous body. This affects its rate of cooling.
3. the composition of the country rock
4. the distance from the igneous body of any particular site in the country rock
5. the volatile content of the magma

Although a lava may well be hot enough to harden the rocks onto which it is erupted, lavas cool relatively quickly and so their metamorphic effects seldom penetrate more

than a few centimetres into the underlying rock. We shall therefore concentrate on the effects of <u>intrusive</u> rocks. The dyke in Fig. 1.2 was probably so near the surface that the confining pressure from the surrounding rocks must have been very low (yet sufficient to prevent the loss of carbon dioxide produced from the heated Chalk (see p. 6). This is not quite the case when we come to consider the emplacement of plutonic masses such as stocks and batholiths, which must have been intruded at depths of at least several kilometres. Nonetheless, the abnormally high temperature gradient produced in the country rock by the nearby magma seems to be of much greater influence than the pressure. This is demonstrated by the field for contact metamorphism on the pressure/temperature graph (Fig. 2.6).

We shall consider the effects of contact metamorphism under several headings, dependent mainly on the composition of the original rock.

Contact Metamorphism of Limestone

Much of the Chalk is a relatively pure limestone and we have already noted the field relations of a dyke cutting Chalk (Fig. 1.2). Although it is not visible on the photograph, the Chalk near the dyke is turned to marble. The same reaction takes place in other types of limestone.

In hand specimens, a marble typically displays a crystalline sugary texture, often streaked with colour bands from traces of impurities. Fossils are usually destroyed, or if marbling is less severe, can be faintly seen as ghosts with ill-defined margins. As with all rocks, a higher degree of magnification is needed to reveal the detail of mineral content and the texture (i.e. the sizes of the grains and the relationship between them). A hand lens is useful in this instance but much more information can be gained by cutting a slice of the rock and grinding it to a standard thickness where most of the minerals become transparent. The resultant thin section is examined beneath a petrological microscope. This is equipped with two pieces of Polaroid arranged so that the thin section may be viewed in plane polarised light or under crossed polars, where the directions of polarisation of the two pieces are at right angles.

Figure 3.1 is a magnified thin section of a pure marble, compared with an unmetamorphosed

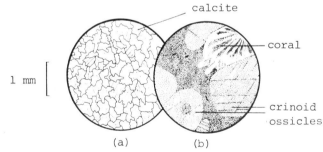

Fig. 3.1 Thin section of marble compared with crystalline limestone: a) marble from Carrara, Italy b) crystalline limestone of Carboniferous age from Derbyshire.

limestone. The marble consists of a granular mosaic of calcite crystals which are tightly interlocking, like the pieces of a jigsaw puzzle. The grains are of uniform size and there is no space between them. By contrast, the unaltered limestone comprises large and small fragments of calcite, embedded in a calcite cement. Several of the larger grains bear clear evidence of their organic origins: in this case this includes two different sections of crinoid ossicles and a coral fragment.

In some field situations, contact metamorphism may be used as an aid to understanding the geological history and structure of the locality. For example, it may be difficult to decide whether a sheet-like igneous body is a lava flow or a sill. The value of contact metamorphism in such cases is this: a lava flow can only affect the rocks beneath it, since it will be cold before renewed deposition can take place on the upper surface. By contrast, a sill is intruded into rocks which are already consolidated and the heat will affect strata lying above as well as below the igneous body.

Figure 3.2 consists of photographs of the top of a dolerite sill intruding Carboniferous limestone in North Derbyshire. This is typical of many situations where exposure is limited. In this case the remains of an old mill dam provide the first hint that an impermeable rock is present. The spheroidally weathered basic rock of the igneous body is evident in the bank but little is then exposed until the small limestone quarry several metres higher. The limestone in the

old mill pond, now dry   dam wall   water flowing from base of old dam wall   marbled limestone beneath grass   fine basalt beneath grass

marbled limestone   spheroidally weathered dolerite

Fig. 3.2 A sill in North Derbyshire.

lowest 1.5 metres of the section is marbled with ghost remains of crinoid stems. A careful search on the intervening grassy bank reveals a fine basalt, which must represent the chilled upper margin of the sill. The precise contact would only be located by digging a trench, but there is certainly enough evidence to show that the igneous body is a sill dipping in conformity with the main mass of the limestone towards the left of the picture. The base of the sill is not exposed.

It is quite common for limestones to have been at least partially dolomitised during their diagenesis. This introduces another element, magnesium, into the system and results in a slightly more complex sequence. The metamorphic reaction at high temperatures is:

$$CaMg(CO_3)_2 \rightleftharpoons CaCO_3 + MgO + CO_2$$

dolomite $\rightleftharpoons$ calcite + periclase + vapour

In this reaction the CO_2 is not so readily confined by pressure and escapes. The periclase usually becomes hydrated in groundwater to brucite ($Mg(OH)_2$).

Many dolomites also contain appreciable quantities of silica in the form of nodules or bands of chert. Under the intense heat of contact metamophism a series of reactions may take place between the silica and the calcite or dolomite. One common reaction may be expressed as follows:

3 dolomite +   1 talc + 3 calcite
4 quartz + water   + 3 carbon dioxide

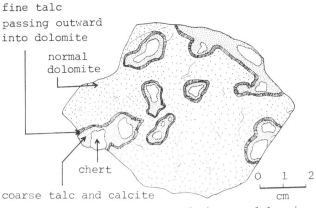

fine talc
passing outward
into dolomite

normal
dolomite

chert

coarse talc and calcite

Fig. 3.3 Reaction rims around chert nodules in a dolomite rock. Aureole of the Benn an Dubhaich granite, Isle of Skye (after Tilley, 1949).

The first indication that the equilibrium of the country rock has been upset is the presence of reaction rims around the chert nodules where they touch the surrounding dolomite. Figure 3.3 is a sketch of a hand specimen of dolomite which exhibits reaction rims around each of the chert nodules contained within it. Two or three narrow zones can be discerned around each nodule, representing various degrees of completion of the reaction. As the junction with the intrusion is approached, other new calcium/magnesium silicate minerals appear in a progressive sequence. Quite commonly, the end member next to an intrusion is wollastonite, whose stability field was discussed on p. 7. A typical progressive sequence is shown later in Fig. 3.11.

Contact Metamorphism of Sandstone

Some sandstones, known as orthoquartzites, consist largely of one component, silica. Thermal metamorphism of an orthoquartzite results in a rock similar in texture to a marble. Recrystallisation occurs until the rock consists of a simple mosaic of interlocking quartz crystals (Fig. 3.4b). The rock is often called a metaquartzite, to distinguish it from its sedimentary equivalent. Comparison with Fig. 3.4a will illustrate the

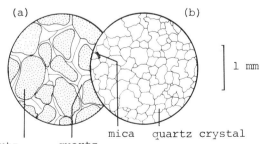

(a)   (b)

1 mm

mica   quartz crystal

quartz   quartz
sand grain overgrowth

Fig. 3.4 Thin sections of quartzites: a) orthoquartzite from the Permian of the Penrith area b) metaquartzite from the aureole of the Foxdale Granite, Isle of Man.

contrast between the two types of quartzite. In the sedimentary rock, the rounded outlines of the sand grains can be clearly seen, cemented together by secondary quartz which

Fig. 3.5a Basic dyke cutting Permian sandstone, Brodick Bay, Isle of Arran.

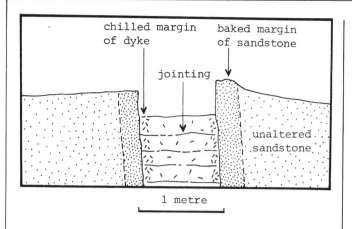

Fig. 3.5b *Sketch section of the dyke in Fig. 3.5a.*

filled the pore spaces some time after the initial deposition. This contrasts strongly with the granular appearance and lack of evidence of former pore space in the meta-quartzite.

In the hand specimen, a metaquartzite dis-plays the same sugary texture as a marble but may be distinguished from it by its greater strength and negative reaction to dilute hydrochloric acid.

Figure 3.5a is a photograph of a basic dyke about 1 m wide cutting Permian sandstones on the Isle of Arran. The sandstone is a moder-ately pure quartz sand, reddened with iron oxides, but within a few centimetres of the dyke it has been 'bleached' by the intrusion. Although not completely recrystallised to metaquartzite, some new growth of quartz has taken place and the baked margins of the sandstone are more resistant to erosion than the dyke itself and therefore stand out (Fig. 3.5b).

An increase in impurities in a sandstone leads to the growth of some new minerals, but the variety is not as great as in the impure limestones until the admixture of clay minerals begins to form a significant pro-portion of the rock. The metamorphism of clay-rich rocks is dealt with in the next section.

Contact Metamorphism of Pelitic Rocks

Sedimentary rocks of fine grain size (less than 1/16 mm) are known collectively as argillaceous, or pelitic rocks. They include clays, shales, mudstones and fine volcanic ashes and are composed primarily of a variety of clay minerals, fine quartz particles and a cement, usually of iron compounds or calcium carbonate. Their bulk chemistry is thus more varied than that of limestones and sandstones and the range of possible new minerals which can develop under conditions of contact metamorphism is considerably more exciting!

As before, igneous intrusions of all sizes may cause metamorphism of surrounding pelitic rocks. The effects are most marked, however, around some of the major plutonic intrusions, where it is often possible to demonstrate progressively more intense metamorphism with proximity to the intrusion. This is a feature of pelitic rocks which is not really shared by the pure limestones and sandstones which are so uniform in composition that they show few progressive changes beyond a slightly coarser grain size nearer to the source of heat.

Progressive metamorphism around the St Austell Granite

We shall concentrate on the granites of Devon and Cornwall, with particular reference to the St Austell granite and the neighbouring Bodmin Moor granite. The granites crop out as several isolated stocks, but we know from gravity survey evidence that they are linked below ground to form an enormous intrusion, known as a batholith (Fig. 3.6). The size of

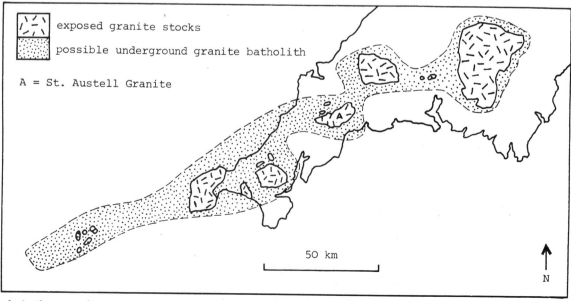

Fig. 3.6 *The granites of Devon and Cornwall, showing possible underground connections.*

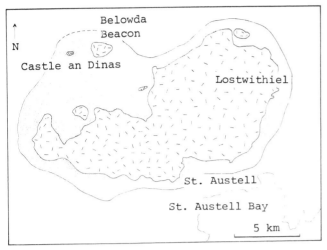

☐ Rocks of Devonian age unaffected by granite intrusion

⟨⟩ Granite

☐ Metamorphic aureole of the St. Austell granite. The closeness of the dots is a rough indication of the intensity of contact metamorphism.

Fig. 3.7 Sketch map of the St Austell granite and its metamorphic aureole, Cornwall.

the batholith is such that it must have exerted a considerable heating effect upon the country rock and taken millions of years to cool.

The map of the St Austell granite stock shows the extent of the outcrop (Fig. 3.7). The granite, which is of Hercynian date, (about 290 million years old) was intruded into Devonian rocks. Most of these were originally shales, siltstones and sandstones, some of which contained a significant proportion of calcium carbonate. For the most part, the Devonian rocks had been tightly folded and affected by low-grade regional metamorphism <u>before</u> the granite was intruded, so much of the country rock would already have been altered to cleaved slate (see p. 24). Nonetheless, the heat of the granite magma had a dramatic effect on the surrounding rocks. The map shows the limit of the affected area, known as the <u>metamorphic aureole</u>. Mostly, it extends some 1 to 2 km from the exposed granite margin but in the north-west it also encompasses the two outlying outcrops of Castle an Dinas and Belowda Beacon. This provides independent evidence in addition to the local gravity results that there is an underlying granite ridge connection with the main mass.

Naturally, the intensity of metamorphism is largely dependent on the distance from the granite margin and there are many localities within the aureole which may be used to demonstrate the various stages through which pelitic rocks pass as they are heated. Working inwards from the unaltered country rock, the usual progression is as follows:

Spotted slate

Often, the first sign of alteration by heat is the development of spots in the country rock. The spots are up to 5 mm across and

have ill-defined edges. They usually consist of clusters of minerals which compose the rest of the rock, only their proportions are different in the clusters. The minerals include quartz, micas, chlorite and iron oxides. The heat was evidently not sufficient to destroy the existing cleavage of the slate.

Andalusite slate

Closer to the granite, the heat must have been greater, and the country rocks become generally more crystalline than in the spotted rock zone. The cleavage of the slates is largely destroyed and new minerals begin to appear. These crystallise with a completely random orientation and the most notable of them is the polymorph of Al_2SiO_5, andalusite (see p. 6). This usually appears as whitish spots which are only identifiable as andalusite beneath the microscope. Occasionally, however, the mineral crystallises in the variety known as chiastolite, named after the cross-shaped arrangement of minute inclusions which sometimes develop within it. Figure

Fig. 3.8 Chiastolite slate.

3.8 is of a chiastolite-slate (though not from Cornwall) which clearly displays the random orientation of the crystals, so characteristic of most contact metamorphosed rocks. When large crystals such as these occur in a finer-grained groundmass in a metamorphic rock, the texture of the rock is referred to as <u>porphyroblastic.</u>

Hornfels

The most intense effects of heat are found in the country rock next to the granite margin or in inclusions of country rock which were engulfed in the granite magma but which did not melt. Such inclusions are known as <u>xenoliths.</u> Recrystallisation in such situations is usually complete and the new crystals interlock in a tight criss-cross mosaic, forming an extremely tough rock which often breaks with a conchoidal fracture, rather like flint. Such a rock is known as a hornfels. Most hornfelses are dark coloured, fine-to-medium-grained rocks and show no evidence of the earlier cleavage of the slates from which they were produced. Sometimes, however, faint traces of original bedding may be preserved, in cases where sediments of coarser grain-size were interbedded with the slates.

Many new minerals are developed in a hornfels, and minerals such as biotite, andalu-

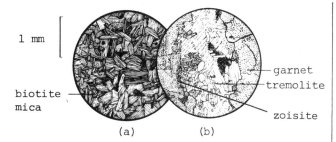

1 mm

biotite
mica

garnet
tremolite

zoisite

(a)   (b)

Fig. 3.9 Thin sections of hornfelses from the
metamorphic aureole of the Bodmin Moor granite,
Cornwall: a) biotite-hornfels showing decussate
texture b) 'calc-flinta' with garnet.

site and cordierite (an Mg, Fe silicate) are
common constituents of the hornfelses of the
inner aureole of the St Austell granite.
Fig. 3.9a is a thin section of a biotite-
hornfels from the aureole of the neighbouring
Bodmin Moor granite and it displays the
characteristic, random criss-cross arrangement
of the crystals known as <u>decussate</u> texture.

In places, the original country rock was a
calcareous siltstone rather than the largely
non-calcareous shales considered above. The
presence of rather more elements in the rock
leads to more variety in the suites of
metamorphic minerals. Near to the granites
these rocks have become pale grey, splintery
rocks known as calc-flintas and they contain
minerals ranging from amphiboles and
pyroxenes to garnet. (Fig. 3.9b, from the
Bodmin Moor aureole)

The progression of changes described above
is, of necessity, rather a general one, and
there are many exceptions. For example, it is
not uncommon to find country rocks which have
remained virtually unaltered, even though
they lie in close proximity to the igneous
contact. Nevertheless, it provides a useful
model for the thermal metamorphism around
many other igneous bodies. Figure 3.10 is a
summary diagram of the most significant
changes which take place during such metamorphic activity.

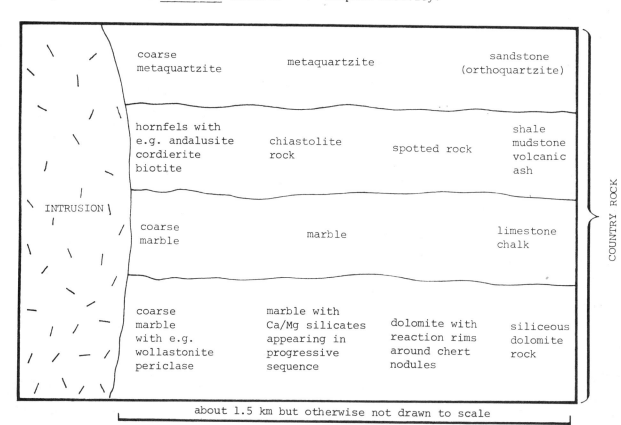

Fig. 3.10 Generalised diagram showing the effects of contact metamorphism.

Metasomatism

So far, we have regarded the metamorphic
reactions within the contact aureole as being
of an <u>isochemical</u> nature, where the only
elements available to make new minerals were
already present in the country rock. It
would be surprising, however, if a rising
magma contributed nothing but heat to its
surroundings. Granite magmas in particular
may be rich in volatile constituents, such
as boron, fluorine and water vapour. Where
the magma encounters permeable rocks or
strata which are fractured or jointed, these

volatiles may escape into the country rock.

Intrusion into limestones may result in
decarbonation and the release of carbon
dioxide which becomes another active
volatile. Reaction between the volatiles and
the rocks of the aureole may then occur and
a range of quite different new minerals may
be formed. Such a process is known as metasomatism. Like the heat of the intrusion
itself, the effects are most marked nearest
to the contact but they may also extend for

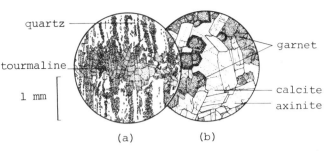

quartz
tourmaline
1 mm
garnet
calcite
axinite

(a)   (b)

Fig. 3.11 Thin sections of rocks of metasomatic origin from the aureole of the St Austell Granite: a) coarse tourmaline-slate b) an axinite-bearing rock.

considerable distances beyond it. In Cornwall for example, there are well-known veins of metal ores which extend several kilometres into the country rock and also traverse parts of the granite masses. Although the ores mostly owe their origin to other causes, it is quite probable that metasomatism had a part to play too.

Figure 3.11a shows a coarse tourmaline-bearing slate from the aureole of the St Austell granite. The big crystals running from right to left show where a crack in the original slate permitted the access of the gases from the granite. Molecular diffusion between the gases and the solid rock produced these and the other tourmaline crystals, (running vertically) thus emphasising the original bedding of the rock.

In some of the calcareous rocks of a contact aureole the variety of new minerals may be very great. Figure 3.11b is just one example from the St. Austell aureole containing axinite (a boro-silicate with calcium, aluminium, iron and manganese) and garnet (a silicate of variable composition, in this case containing calcium and iron). Figure 3.12 shows, in general terms, the type of

situation which most favours metasomatism, and where metasomatic ore minerals are likely to be found.

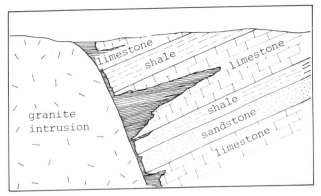

Fig. 3.12 Metasomatism around a granitic intrusion, showing the more wide-reaching effects in the permeable limestones.

Basic magmas are mostly injected at a higher temperature than acid ones and so they might be expected to exert a wider influence than granites. In practice, however, the contact aureoles around gabbros are usually narrower than around granites of comparable size. There is also a wide discrepancy between granites themselves and the relationship between the size of the pluton and the width of the aureole is by no means as clear-cut as might have been implied in an earlier section (p. 9). It is well known that gabbro intrusions show less evidence of volatile components than granites such as the St Austell mass. This also applies to many granites with narrow contact aureoles, such as the Isle of Arran granite whose aureole has a maximum width of 180 m. It is now becoming increasingly apparent that not all the heat is transferred by conduction through the country rock. At least a proportion of it is conveyed by the medium of late-stage aqueous fluids.

Relict Textures and Structures in Contact Aureoles

One of the most interesting problems for the metamorphic petrologist is to try to determine the nature of the rock before it became metamorphosed. In a way, it is rather like a detective trying to find the identity of a document from a charred piece recovered from a bonfire. In the outer zones of a metamorphic aureole the job is not too difficult but in the hornfels zone there may be few clues remaining. It must be remembered that, in addition to the whole range of sedimentary rocks, older igneous rocks and existing regionally metamorphosed rocks may also be altered by the heat of a younger intrusion.

In the case of sedimentary rocks, bedding structures are quite commonly preserved, especially when the rock was not uniformly pelitic but contained layers of material of coarser grain size. Figure 3.13 shows a contact between the Skiddaw granite and the crystalline hornfelses produced from sedimentary rocks of the Skiddaw Group in the

Fig. 3.13 Contact between Skiddaw granite (light) and dark hornfelsed rocks of the Skiddaw Group, Cumbria. (Walking stick is 90 cm long).

Lake District. Near the contact, superb examples of relict folding may be found in the hornfelses (Fig. 3.14).

Fig. 3.14 Relict folding in crystalline hornfels. Loose block a few metres from the locality in Fig. 3.13.

Elsewhere in the same aureole, where the hornfelsing was less complete, there is microscopic evidence of the former cleavage of the pelitic country rocks. Figure 3.15 shows that the crystals of cordierite contain many tiny inclusions of other minerals, a phenomenon known as poikiloblastic or sieve texture. These include platy minerals such as micas, which have a very marked preferred orientation (from left to right in Fig. 3.15).

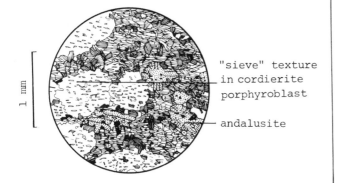

"sieve" texture in cordierite porphyroblast

andalusite

1 mm

Fig. 3.15 Thin section of a cordierite-andalusite slate from the aureole of the Skiddaw granite, Cumbria.

This fabric is not the result of the heat of the granite several hundred metres away, but marks the direction of the existing cleavage of the slate, produced by regional stresses earlier in the tectonic history of the Lake District. The fact that the cleavage pre-dates the intrusion is of great importance to structural geologists trying to elucidate the geological history of the area.

Similarly, some of the textures or structures of altered igneous rocks may be preserved after their metamorphism. Figure 3.16 is of a hornfels from the aureole of the Shap granite in Cumbria. It was once an amygdaloidal lava, i.e. one where gas holes had been filled in by later minerals such as calcite or zeolites. Although the lava is now a tough flinty hornfels, the amygdales are still clearly visible, albeit now filled with a recrystallised version of their original minerals.

Fig. 3.16 Hornfelsed amygdaloidal lava from the aureole of the Shap granite, Cumbria.

These are just a few of the relict textures and structures which may be seen in contact metamorphism. Similar features occur in regional metamorphism and some are found in both regimes. A further description of such phenomena will be found in Section 5.

4
Dynamic Metamorphism

In contact metamorphism we have been considering the effects of the increase in temperature brought about by the nearby intrusion of an igneous body. Although at the same time, stresses must be set up in the surrounding rocks, their metamorphic effects are secondary to those of the heat. By contrast, in areas where major faults have acted, rocks lying along the fault plane are often intensely sheared and altered. Heat is then usually of less importance as a metamorphic agent although friction may be sufficiently intense to produce local melting and the injection of mobile material (known as pseudotachylite) a short distance into the surrounding rocks.

Faulting may occur at many different levels within the crust. Generally, faulting at the higher levels occurs where the rocks are most brittle. Rocks lying along the fault plane are shattered and are frequently re-cemented by solutions carried in ground water to form a fault-breccia (Fig. 4.1). In pelitic rocks, or in situations where faulting results in the broken material being ground down very finely, the end-product may be a fine clayish material known as gouge. There is disagreement among geologists as to whether fault-breccia and gouge represent true metamorphic rocks.

Mylonites

There is no doubt, however, about the metamorphic nature of rocks produced along major fault planes at greater depths in the earth's crust. Here, both the ambient temperature and the pressure of overlying rock are higher, and the outcome of movement along the fault plane is more dramatic. It is characteristic for broken rock material to be drawn out into lens-shaped fragments, usually most evident

Fig. 4.1 Fault-breccia cutting thin banded Moine gneisses, Glen Garry, Perthshire.

on a microscopic scale. These are embedded in a finer-grained matrix composed of more minutely ground particles, sometimes accompanied by recrystallisation and the formation of new minerals. The process has been likened to the way in which corn is ground into flour by being crushed between two heavy rotating millstones and the name mylonite has been applied to rocks of this type, based on the Greek word for a mill.

The Glarus Nappe, Switzerland

Mylonites are best developed along major thrust planes and the section in Fig. 4.2 is of the Glarus Thrust in the Alps, where the signficance of such faulting was first fully appreciated. Movement along the thrust has been so great that a huge sheet of rock

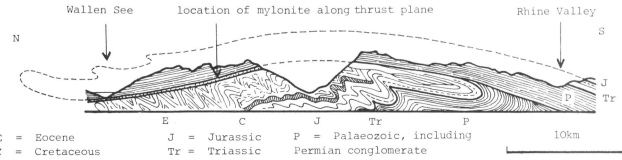

E = Eocene   J = Jurassic   P = Palaeozoic, including
C = Cretaceous   Tr = Triassic   Permian conglomerate

Fig. 4.2 Section showing the structure of the Glarus Nappe, Switzerland.

has been moved bodily some 35 km from the site of its deposition to lie above strata which are of younger age. Such sheets are known as nappes. Along the thrust plane of the Glarus Nappe some superb mylonites are developed and the thin section shown in Fig. 4.3 is of the Lochseiten Mylonite, formed largely of marbled limestone fragments. The larger particles have been pulled out into the characteristic lens shapes which lie in a sub-parallel fashion, imparting a rough layering, or foliation to the rock, (running approximately from left to right in the thin section). These large particles exhibit the effects of strain when viewed in cross-

polarised light beneath the microscope. In some parts of the thin section, notably the top right, calcite grains occur in a close mosaic, making an angle of about 120° where they meet. The texture of this kind of mosaic is referred to as granoblastic. The grains do not show evidence of strain and it is clear that they formed by recrystallisation relatively late in the history of thrust movement. Figure 4.4 is a generalised, detailed section of the thrust zone showing the relationship between the much-travelled nappe above and the stable, unmoved mass (autochthon) beneath. The mylonite is about 1 m thick.

It is always interesting to know how much rock has been moved and at what rate. Stratigraphical considerations suggest that there might have been some 2.5 km of material lying above the thrust plane. Making certain rather sweeping assumptions about mineral stability and rates of reaction, it has been calculated that the thrust travelled at a minimum rate of 3.5 mm per year. Assuming this rate, you can easily calculate the maximum time it would have taken for the nappe to travel the estimated distance from its origins 35 km away.

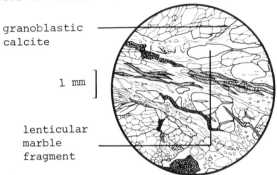

granoblastic calcite

1 mm

lenticular marble fragment

Fig. 4.3 Thin section of Lochseiten Mylonite, Swanden, Canton Glarus, Switzerland.

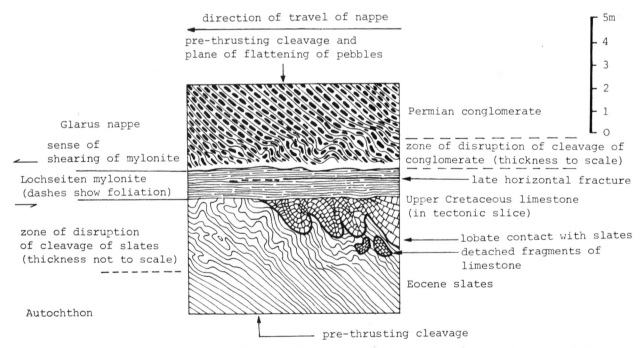

direction of travel of nappe

pre-thrusting cleavage and plane of flattening of pebbles

5m
4
3
2
1
0

Glarus nappe

Permian conglomerate

sense of shearing of mylonite

zone of disruption of cleavage of conglomerate (thickness to scale)

Lochseiten mylonite (dashes show foliation)

late horizontal fracture

Upper Cretaceous limestone (in tectonic slice)

zone of disruption of cleavage of slates (thickness not to scale)

lobate contact with slates

detached fragments of limestone

Eocene slates

Autochthon

pre-thrusting cleavage

Fig. 4.4 Generalised cross section through the Lochseiten Mylonite zone at the base of the Glarus Nappe. (From Schmid, 1976)

The Moine Thrust, Scotland

Within the North-West Highlands of Scotland there exists an equally famous example of large-scale thrusting belonging to an earlier phase of movements known as the Caledonian Orogeny. A cursory glance at the map of part of the area (Fig. 4.5) would suggest that a major unconformity exists

between the Moine Schists and the underlying rocks. There is, however, clear evidence that the junction is a thrust and not an unconformity, since the Moine Schists are of Precambrian age, yet rest on uninverted orthoquartzites and limestones formed in the Cambrian Period (Fig. 4.6). In addition, most of the Moine Series rocks have been metamorphosed by regional processes and yet lie upon unaltered sedimentary rocks. The tectonics of the Moine Thrust itself and associated thrusts and reversed faults in the rocks beneath are consistent with the derivation of the Moine Schists from an original site probably many kilometres to the east-south-east. The Moine Thrust has therefore produced a nappe, similar in many

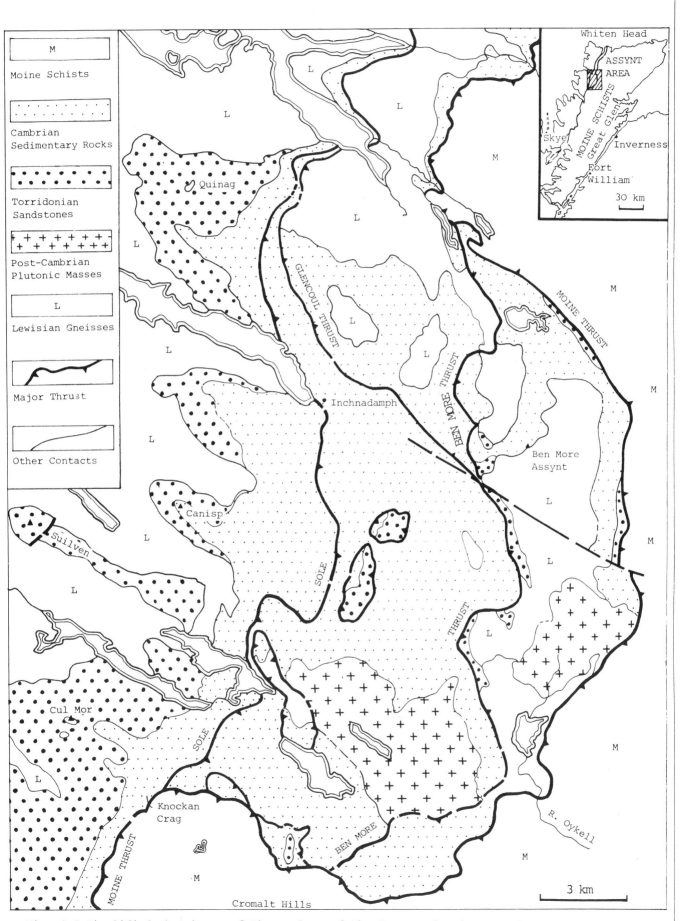

Fig. 4.5 Simplified sketch map of the geology of the Assynt District, North West Highlands of Scotland.

Fig. 4.6 The Moine Thrust plane at Knockan Crag, Moine Schists (dark) have been thrust over the light-coloured Cambrian limestones.

respects to the Glarus Nappe of the Alps (Fig. 4.7). Finally, mylonite is present at many localities along the plane of the Moine Thrust, and indeed along some of the other thrust planes in the area.

The hand specimen in Fig. 4.8 is a typical mylonite from the Moine Thrust plane. It shows well the fine-grained, rather flinty texture of the rock and the closely spaced planes produced by the alignment of its constituent grains.

Microscope work shows that there is considerable variation among mylonites, depending upon the composition of the rock caught up in

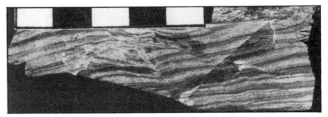

Fig. 4.8 Mylonite, Moine Thrust plane, Assynt district.

the 'milling' process along the thrust plane. Figure 4.9 is of a fragment of Lewisian Gneiss in the early stages of mylonitisation.

1 mm

Fig. 4.9 Thin section of mylonitised Lewisian Gneiss.

Unlike the Lochseiten mylonite which was produced from limestone, there has been little recrystallisation in this specimen, the matrix consisting of fragments of the original rock which have simply been ground down more finely than the large lens-shaped particles.

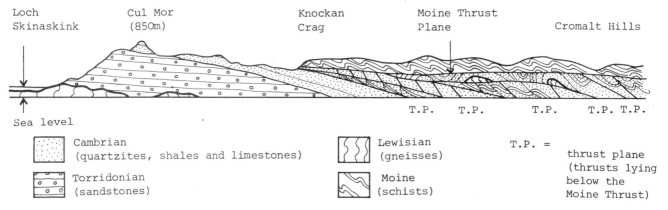

N.W.   S.E.

Loch   Cul Mor   Knockan   Moine Thrust
Skinaskink   (850m)   Crag   Plane   Cromalt Hills

Sea level

T.P.   T.P.   T.P.   T.P. T.P.

Cambrian
(quartzites, shales and limestones)

Torridonian
(sandstones)

Lewisian
(gneisses)

Moine
(schists)

T.P. = thrust plane
(thrusts lying
below the
Moine Thrust)

Fig. 4.7 Diagrammatic section of part of the Moine Thrust zone. Length of section is 7 km.

5

Regional Metamorphism

In contrast to the relatively restricted occurrences of rocks of contact or dynamic metamorphic origin, there are many areas of the world where hundreds of square kilometres of country are occupied by metamorphic rocks which owe their origin to widespread processes of regional metamorphism. Examples include large parts of the Precambrian shield areas, like the Baltic Shield, and former orogenic belts such as the mountains of Norway and Sweden. As in the other types, regional metamorphism commonly results in profound changes in the mineralogy, texture and structure of the original rocks. We shall again assume that most of the transformations are isochemical ones, where the bulk chemical composition of the rock has remained much the same, apart from the escape of water and other volatiles. As before, these fluids play an important part in the transfer of ions within the solid rock and thus assist in the production of new minerals. The nature of the regionally metamorphosed rock thus depends quite strongly on the original rock

type. Some end products can easily be related to their sedimentary or igneous parent; in others this is virtually impossible.

In contact metamorphism we regarded change of temperature as the dominant factor and in dynamic metamorphism the effects were attributed mainly to locally intensified pressure. A glance back to Fig. 2.6 shows that regional metamorphism and its sub-division of burial metamorphism cover a much wider range of temperatures and pressures, which are usually only encountered at considerable depths in the earth's crust or upper mantle.

Figure 5.1 provides a more quantitative view of some mineral transformations which are of importance in regional metamorphism, although allowance must be made for the discrepancies between the results of different workers (noted on p. 6). Of the stability fields we have discussed before, it is interesting to compare the relative positions of the albite/

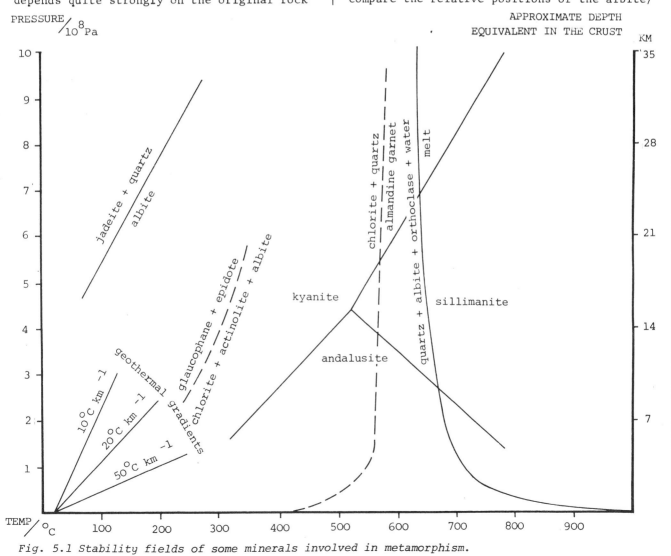

Fig. 5.1 Stability fields of some minerals involved in metamorphism.

jadeite stability field and that of the triple point of the polymorphs of Al_2SiO_5. Clearly, the main requirement for jadeite to be produced in a rock is that the pressure should have been high, usually achieved in regions of low heat-flow combined with intense pressure or by great depth of burial in the crust. Rather hotter, but otherwise similar conditions are needed before the minerals glaucophane and epidote will become stable, although the precise position of the stability field is also very sensitive to rock composition. Glaucophane is a deep blue colour and so rocks in which it is abundant are commonly known as 'blueschists'. Together with jadeite-bearing rocks and others, these comprise the results of low temperature, high pressure metamorphism or of burial metamorphism. The tectonic significance of these rocks will be discussed later.

By contrast, greatly raised temperatures are necessary for andalusite, kyanite and sillimanite to become stable. Of these three, we have already seen that andalusite is common in hornfelses around igneous plutons. Sillimanite, too, may occur in high-grade contact metamorphosed rocks, but for kyanite to form, considerable pressure must also have been acting, usually from a combination of directed stress and hydrostatic pressure.

The approximate position of another significant stability field is shown, namely where the iron-garnet, almandine first appears. We have already seen (p. 14) that some varieties of garnet occur in contact aureoles, but the mineral is more typical of regional metamorphism. In both cases the incoming of garnet depends very much on the composition of the parent rock.

It must again be emphasised that metamorphism, by definition, involves changes which take place in the solid state. However, under conditions of very high temperatures and pressures it is possible for rocks to begin to melt and there is thus a very close relationship between regional metamorphism and the origin of some igneous rocks. The graph includes a melting point curve for a 'wet' granitic mixture (quartz + albite + orthoclase + water) and it is clearly within the same part of the pressure/temperature field as the more extreme conditions experienced in regional metamorphism.

Figure 5.1 also shows several geothermal gradients. This term describes the rate at which temperature increases with depth in the earth. It is derived partly by direct measurement in deep mines and boreholes and is known to vary widely. In Great Britain, for example, it varies between $13^{\circ}C$ km^{-1} and 45° km^{-1}. Possible geothermal gradients below the topmost few kilometres are derived by calculation, based on a variety of data, including laboratory measurements on the thermal properties of rocks.

It is clear from the graph (Fig. 5.1) that burial metamorphism occurs where the geothermal gradient is low and thermal metamorphism where it is very high (on account of the nearby injection of magma). However, regional metamorphism can occur over a range of geothermal gradients. The higher grade end of regional metamorphism probably represents an abnormally high thermal gradient, perhaps raised by a convective current from the mantle, frictional forces, or an unusual degree of depression of crustal rocks into a higher temperature regime.

Because of the isochemical nature of most regional metamorphism, the nature of the parent rock imposes constraints on the possible product, so as with contact metamorphism, we shall treat the subject under headings based on the original rock type.

Regional Metamorphism of Limestone

The effects of regional metamorphism of a pure limestone may be dismissed very briefly; a marble is produced which differs very little from similar rocks in contact aureoles! Both consist of interlocking (granoblastic) grains which produce a sugary texture in the hand specimen and which re-

Fig. 5.2 Flow folds in plastic metamorphosed limestone (after Balk).

semble the thin section in Fig. 3.1, (which is actually a regionally metamorphosed limestone, not a contact marble). The microscope reveals that marbles which formed under moderate stress conditions contain bent and twinned crystals. Under higher pressures and temperatures the grain size of the marble becomes coarser. On a larger scale, the outcome of plastic flow may be seen in some exposures of marble (Fig. 5.2).

As in contact metamorphism, impure limestones, especially dolomites, may show the development of a wide range of new minerals. Many of these are calcium-rich silicates which appear in a well-documented sequence, providing evidence of progressively more intense conditions of metamorphism. Such considerations are, however, beyond the scope of this Unit.

Regional Metamorphism of Sandstone

In the case of a nearly pure quartz sandstone, regional metamorphism produces a very similar result to contact processes. A metaquartzite is formed, with the same tightly interlocking mosaic of quartz crystals (See Fig. 3.4). Increasing metamorphic intensity again often leads to a coarsening of grain size. Beneath the microscope, the quartz crystals may show the effects of strain, with rows of tiny fluid-filled pores crossing each grain in the direction of stress, but otherwise there is little difference from the metaquartzites produced by contact metamorphism.

Fig. 5.3 Metaquartzite, showing cross-bedding remaining from the original sediment. Glencoe, Scotland.

Because of the relatively rigid nature of quartz grains, primary sedimentary structures are frequently preserved, becoming distorted only at the highest degrees of metamorphism. Figure 5.3 shows a cross-bedded metaquartzite from the Dalradian of Glencoe. Although the rocks of the area have been subjected to moderately high pressures and temperatures, the original bedding is still clearly visible. Such a relict structure when found in situ may be of considerable importance, not only in helping to determine the former sedimentary environment but also as an indicator of the 'way-up' of the sequence. In Fig. 5.3 the specimen is the right way up, since the cross-bedding planes taper towards the bottom of the picture and are truncated towards the top (just below the scale marker).

Material other than quartz in a sandstone provides more opportunity for new minerals to be formed during metamorphism. As in contact metamorphism, the variety of possible new materials increases with increasing proportions of clay minerals, micas and feldspars in the original sandstone. The range of possibilities is again wider than we need consider here, although some idea can be gained from the next section which examines pelitic rocks.

Regional Metamorphism of Pelitic Rocks

Since we are primarily dealing with isochemical changes, pelitic rocks, with their wide range of finely divided minerals, show some of the greatest variety of changes when they are metamorphosed. Just as in contact aureoles, pelitic rocks which have undergone regional metamorphism frequently exhibit evidence of progressive changes reflecting increasing temperature and pressure during metamorphism.

Rocks which have been affected by increased pressures and only moderate temperatures are generally thought of as low-grade metamorphic rocks. At increased temperatures, changes in the rocks become more profound and the new rocks are regarded as high-grade. Increasing

metamorphic grade results in changes to all aspects of the rock, including its mineralogy, texture and fabric and to the style of deformation in which it is involved. The bulk chemical composition is assumed to remain much the same, apart from volatiles, although there is evidence to suggest that metasomatism, i.e. the introduction of new material, becomes important in the highest grade material.

In simple terms, the pelitic rocks show the following progressive sequence from low to high grade:

pelitic sedimentary rock → slate → phyllite → schist → gneiss.

TABLE 5.1 THE CHARACTERISTICS OF REGIONALLY METAMORPHOSED PELITIC ROCKS.

	TEXTURE	COLOUR	MINERAL CONTENT	ROCK NAME	ORIGINAL ROCK
obvious sub-parallel alignment	Very fine-grained	Generally dark grey, may be green or red.	Fine or very fine-grained mica flakes in parallel orientation, often visible in hand specimen. Also very fine-grained quartz, feldspar, etc., generally only identifiable in thin section.	SLATE	Usually shale or volcanic ash.
	Fine-grained	Generally medium to dark grey, with micaceous sheen.		PHYLLITE	
	Medium-to coarse-grained, well-developed mineral alignment.	Variable, depending on nature of dominant minerals commonly with micaceous sheen.	Most obvious will be one of: biotite, muscovite, chlorite, hornblende. Quartz and feldspar also usually present, and sometimes other minerals (e.g. calcite, garnet)	SCHIST	Shale, greywacke, volcanic rocks.
	Medium-to coarse-grained, may be banded.	Pale and dark bands in alternation	Generally quartz and feldspars dominant, along with micas, hornblende and other minerals.	GNEISS	As above, also arkose, some plutonic rocks.

Some of the characteristics of each rock are summarised in Table 5.1. These are described at greater length in the section that follows.

Slates

Many slates and shales superficially resemble each other, since each is bounded by parallel planes along which it splits quite readily. Students are often advised to test the hand specimen by holding it up by a corner and tapping it gently: the slate usually emits a more metallic sound than the shale, which responds with a rather uninteresting dull thud. The difference in sound reflects a marked difference in the arrangement and degree of crystallinity of the two rocks. In the shale, the grains consist of finely divided clay minerals and quartz grains, usually cemented together by iron hydroxides. The parallel planes represent lamination planes, produced by the flaky clay minerals having settled out of suspension parallel to each other. There are also usually many pore spaces in the rock. In the slate, on the other hand, the clay minerals have recrystallised into tiny mica flakes which lie parallel to the direction of the least stress acting at the time when the rock was deformed. This is usually at right angles to the direction from which the greatest stress came and often bears little relation to the original bedding. The new direction in which the rock now splits is called the cleavage and Fig. 1.3 showed an example where this cleavage is roughly at right angles to the original bedding, indicated by slight colour contrasts within the specimen.

0.1 mm [

— quartz
— sericite mica and chlorite
— haematite

Fig. 5.4 Highly magnified thin section of slate. North Wales.

Figure 5.4 is a thin-section of a Welsh slate cut at right angles to the cleavage planes. Most of the platy minerals are thus seen edge on and the cleavage is clearly visible running from top to bottom. The main constituents are flakes of sericite mica, chlorite and haematite. The large white fragments are of quartz remaining from the original sediment. Note the high degree of enlargement: even the largest grains are only 0.05 mm in diameter. The slate is virtually non-porous.

This aligned fabric of the mineral constituents of the slate is usually referred to as the slaty cleavage. Slaty cleavage is only one type of such mineral alignment to which the general name of foliation is given. As we shall see, most regionally metamorphosed rocks are foliated, although none of them split quite so regularly as the slates.

Of course, slates do not form from shale alone - other pelitic rocks such as mudstones and siltstones and even fine volcanic ash may produce slates when metamorphosed. The Welsh roofing slates are mostly derived from dark muds, whilst the attractive green slates of the Lake District, used in ornamental work, originated from volcanic ashes.

Although the minerals of a slate are too fine to be seen even with a hand lens, some slates do contain abnormally large crystals, such as the ones in Fig. 5.5. In any metamorphic rock, large crystals such as this, set in a finer grained groundmass are known as porphyroblasts. The porphyroblasts in Fig. 5.5 are of pyrite (FeS_2). The presence of

Fig. 5.5 Porphyroblasts of pyrite in slate. Ballachulish, Scotland.

pyrite in a slate is normally taken to mean that the conditions of deposition of the original mud were reducing. Iron hydroxides are common in most muds and the sulphur was probably produced by bacterial action. Later chemical reactions during metamorphism resulted in the production of the pyrite.

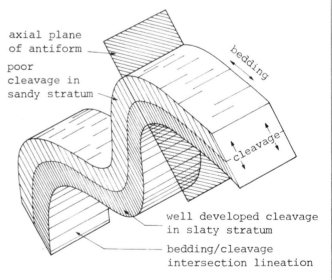

Fig. 5.6 Illustration of the geometrical relationships between axial-plane cleavage and bedding.

It has long been known that there is a close relationship between slaty cleavage and the trends of folds in slates, since both may owe their origin to the same stresses. Fig. 5.6 depicts this relationship and shows that the cleavage is approximately parallel to the axial plane of the fold. For this reason it is often known as axial plane cleavage and is of value in determining the direction of fold axes in areas of poor exposure. Where the cleavage intersects the bedding planes, lines known as lineations, are frequently to be seen. Where pelitic rocks were interbedded with coarser grained sediments, and then metamorphosed, the cleavage is better developed in the slates than in intervening sandstones or quartzites, where the grains are more rigid and do not readily acquire the orientated fabric. Figure 5.7 is of such an

Fig. 5.7 *Folding in interbedded slates and sandstones, Jangye-rin, Cornwall.*

interbedded sequence on the south coast of Cornwall. The tight style of folding with fracturing in many of the crests of the folds is typical of deformation undergone in low grades of regional metamorphism, where stresses are high but temperatures are relatively low. An axial plane cleavage is just visible in the slate horizons but is not developed at all in the more massive beds. The phenomenon is more clearly seen in the close-up photograph of some Silurian rocks in Cumbria (Fig. 5.8). The sequence consists of graded bedded greywackes tilted to some 80° or so. The cleavage, which is vertical, is visible in the slates but not in the intervening, coarser-grained horizons.

Fig. 5.8 *A graded bedded sequence showing slaty cleavage in the finer-grained portions. Shap Fell*

Phyllites

At slightly higher grades of metamorphism, temperatures were presumably higher and recrystallisation was facilitated, resulting in a noticeable increase in grain size. There exists a group of rocks known as phyllites whose grain size is sufficiently coarser than that of a slate for the constituents to be seen with the hand lens. The overall effect is to impart a silvery sheen to the rock, as though it were wet. The mineral content of a phyllite is similar to that of a slate, except that few of the original minerals remain unchanged and there is an increase in mica content.

Like the slate, the minerals of the phyllite have a marked foliation fabric, although it is rather less regular than in the slate and most phyllites are not quite so easy to split thinly for roofing purposes.

Styles of deformation in phyllites are often rather more plastic than in slates. In Fig. 5.9 the early formed cleavage of the phyllite, (parallel to the top of the specimen) has subsequently been folded by stresses acting in a direction at right angles to those which produced the cleavage. This change of direction is not surprising when one recalls that regional metamorphism is associated with widespread mountain-building activity which may continue for many millions of years.

Fig. 5.9 *Folded phyllite, Anglesey.*

Schists

Schists are rocks produced under higher temperatures than slates or phyllites and are correspondingly coarser in grain size, the grains now being visible to the naked eye. As with the other groups of rocks, each category represents a range of conditions and there is considerable diversity within the schist group, both in grain size and mineral content. Most schists are rich in quartz and micas and possess a markedly foliated fabric. The foliation planes are sometimes referred to as planes of schistosity and they are usually wrinkled so that a clean planar cleavage is impossible. Instead, schists tend to split into irregular sheets. There is much debate as to whether the foliation planes represent a further development of an earlier cleavage, or whether they are produced parallel to the bedding planes of the original sediment.

In addition to quartz and micas, a number of new and distinctively metamorphic minerals usually occur in a schist. These are derived mostly from the clay minerals of the original rock and include minerals such as the garnet family, staurolite, kyanite and sillimanite. Feldspars are also frequently produced. The stability fields of some of these minerals have been given already in Fig. 5.1. Andalusite is rare in regionally metamorphosed rocks, since the strong directed stress actually inhibits its growth.

Fig.5.10 Garnet-mica schist, Glen Roy, Scotland.

Schists are quite commonly porphyroblastic, like the garnet-mica schist in Fig. 5.10. It is the custom among metamorphic petrologists to prefix to the rock name the identity of any important mineral which forms porphyroblasts. This is a simple system to follow, although it may result in some frighteningly long names!

Schists are often severely deformed by continuing stresses, the style of deformation becoming increasingly plastic in schists from higher temperature regimes. Sometimes examination of thin sections of porphyroblastic schists enables us to determine the sequence in which deformation and recrystallisation occurred. Fortunately, minerals such as garnet are not restricted in their growth by the intense pressure around them. Indeed, the large size of the garnets in the coarse schist in Fig. 5.11 would suggest that the mineral has a strong force of crystallisation.

Fig. 5.11 Coarse garnet schist, South Orkney Islands.

As an exercise, study the thin section drawings, (Fig. 5.12) which show three garnet schists, each having a rather different history from the others. Assuming that the foliation in each rock is due to directed stress causing recrystallisation of clay minerals into micas, try to match the appropriate description to each diagram.

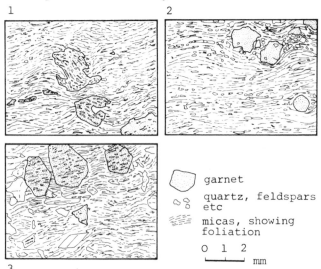

		garnet
		quartz, feldspars etc
		micas, showing foliation

0 1 2
⊢——⊢——⊣ mm

Fig. 5.12 Three garnet schists with different histories of crystallisation.

Descriptions for use with the exercise and Fig. 5.12

a) *The garnet crystals enclose grains of the ground-mass which lie parallel to the foliation, so they grew after the main phase of deformation.*
b) *A 'snowball garnet' has been produced where inclusions have been caught up during growth of the garnet and rolled into part of a spiral. Growth of the garnet took place at the same time as the development of the foliation.*
c) *The foliation is deflected around the garnet porphyroblasts, suggesting that the garnets were produced before the main phase of deformation.*

Answers can be found on page 29

Gneisses

Gneisses are associated with the highest temperatures encountered in regional metamorphism and are often found in close proximity to rocks where partial melting has taken place (see p. 32). Although gneisses can be formed isochemically from pelitic rocks, analyses of many gneisses show that metasomatism may have been an important additional factor.

There is considerable overlap in grain size between a schist and a gneiss: both may be medium-to-coarse-grained. The distinction lies in the nature of the foliation. In contrast to the closely spaced mica-coated planes of the schist, developed in a homogeneous rock, the minerals of the gneiss are segregated into light and dark coloured bands. These may range in width from a millimetre or so to several metres and are usually considerably less regular than the foliation planes in the lower-grade rocks. In the specimen in Fig. 5.13 the gneissose banding shows considerable distortion and has been injected later by cross-cutting quartz lenses.

Fig. 5.13 Gneiss with contorted foliation and later injections of quartz.

The light-coloured minerals of gneisses are predominantly quartz, feldspars and muscovite mica. The dark ones include biotite mica and ferromagnesian minerals such as hornblende. The aluminosilicates found in schists may also occur in gneisses, in which case the name is used as a prefix: e.g. kyanite gneiss, sillimanite gneiss etc. Gneisses often contain porphyroblasts, as do the other categories of metamorphic rocks, but one example deserves particular mention. In some cases, the porphyroblasts are lens-shaped clots of feldspar and are roughly aligned along the foliation. They ·are

Fig. 5.14 Augen gneiss, the Alps.

supposed to resemble eyes, and are given the German name for eyes, <u>augen</u>. Hence the name of augen gneiss for the specimen in Fig. 5.14.

In addition to, or instead of, the gneissose banding mentioned above, gneisses rich in hornblende often exhibit a <u>lineated</u> fabric. Hornblende and other amphiboles commonly crystallise with a needle-like habit, in contrast to the platy habit of the micas. These needles tend to lie with a preferred orientation known as a lineation (Fig. 5.15). Such lineations may indeed occur in schists if needle-like crystals are developed.

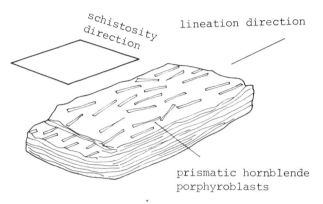

Fig. 5.15 Sketch of a schist showing both lineation and foliation (schistosity), Sulitjelma, Norway.

On the large scale, gneisses show a much more plastic style of deformation than the lower-grade rocks. Fig. 5.16 demonstrates some very convoluted folding in a strongly banded Norwegian gneiss. In places, the folds are discontinuous, for example, near the top of the specimen.

Fig. 5.16 Plastic-style folding in gneiss, Norway.

Regional Metamorphism of Igneous Rocks

Basic igneous rocks are themselves of high-temperature origin and it might be supposed that they would be incapable of being altered any further. However, this is not the case in regional metamorphism, where rocks may be held for a long time under conditions of high stress and high temperature, even though the temperature is still well below that of the original basic magma. Regional meta-morphism of basic igneous rocks seems to have taken place most readily when water or other volatiles have been introduced into the rock

215

during deformation, since most basic rocks are too 'dry' for the necessary reactions to take place.

Basic igneous rocks consist essentially of plagioclase feldspar and ferromagnesian minerals such as the pyroxene, augite. Under regional metamorphism, the pyroxene is converted to amphibole, largely by a process of hydration. The resulting rocks are known as amphibolites, or hornblende gneisses if they are banded.

Similar principles apply to the regional metamorphism of the other categories of igneous rocks. Most notable are the ultra-basic rocks which give rise to serpentinites and soapstones, and acid lavas and granites, from which granitic gneisses may be derived.

It is not easy to ascertain the igneous origin of some of these rocks. Evidence may range from the field relations of the rocks, where distinctive relict structures such as the remains of pillow lava formations may be detected, down to the microscopic scale. Figure 5.17, for example, is interpreted as a relict phenocryst in a former basic lava.

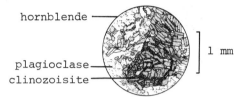

Fig. 5.17 Amphibolite from a porphyritic lava-flow, including relict phenocryst (Sulitjelma, Norway).

Regional Metamorphism in the Scottish Highlands

The progressive nature of regional metamorphism has already been emphasised, notably in the pelitic rocks, where the biggest number of sub-divisions can be established. It remains to be shown that such progressive characteristics can be identified in the field. Whilst many regions have been investigated in the effort to delimit zones of increasingly more intense regional metamorphism, the Grampian Highlands of Scotland still provide one of the best known examples. Parts of the region were mapped by G. Barrow at the turn of the century and his work has subsequently been extended by other geologists.

Fortunately, the bulk of the Dalradian metamorphic rocks which compose the Grampians are derived from pelitic parent rocks, although it is not possible to demonstrate the transition from unmetamorphosed pelitic

rocks into slates etc. In very general terms, there is a progression from the south and west to the north and east, passing from slates, through phyllites and schists to gneisses, many of which show evidence of partial melting (Fig. 5.18). This gives a rough zoning of metamorphic rocks, but Barrow's main contribution was to identify characteristic mineral assemblages which were evidently more sensitive to changes in the conditions of metamorphism and therefore allowed a finer zoning to be carried out.

TABLE 5.2 MINERAL ASSEMBLAGES PRODUCED IN PELITIC ROCKS BY PROGRESSIVELY MORE INTENSE CONDITIONS OF REGIONAL METAMORPHISM

METAMORPHIC ZONE	ASSEMBLAGE OF MINERALS
Chlorite	*chlorite*, muscovite, quartz, (clay minerals)
Biotite	*biotite*, chlorite, muscovite, quartz, sodic plagioclase
Garnet	*garnet*, biotite, muscovite, quartz, sodic plagioclase
Kyanite	*kyanite*, garnet, biotite, muscovite, quartz, sodic plagioclase
Sillimanite	*sillimanite*, garnet, biotite, muscovite, quartz, sodic plagioclase

Table 5.2 lists the mineral assemblages most commonly encountered in the Dalradian pelites of the Grampians. They are divided into five groups marking five zones of metamorphic rocks, arranged in order of increasing grade. It is immediately apparent from Table 5.2 that there is considerable overlap of minerals between the zones; for example, quartz occurs throughout and is therefore of no value for determining differences in the conditions of metamorphism which once affected the sequence. Other minerals, however, occur in several zones, but not all of them. It is possible to use the first appearance of a new mineral, such as biotite, to mark the beginning of a

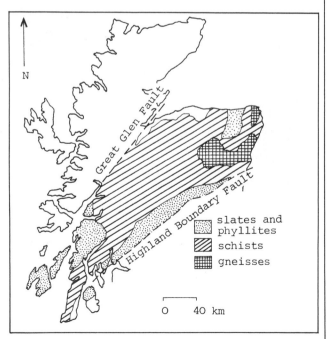

Fig. 5.18 Generalised distribution of regionally metamorphosed rocks in the Grampian Highlands of Scotland.

new zone. The end of that zone is not mapped on the disappearance of the mineral, but on the first occurrence of the <u>next</u> one in the series. For example, the next zone after biotite in order of increasing intensity is marked by the first appearance of garnet crystals, even though biotite is still present, and so on. In the assemblages of Table 5.2 the key, or <u>index</u> mineral for each zone is given in italics and the zone named after it, although the use of garnet as an index may be criticised for the reasons given on p. 22

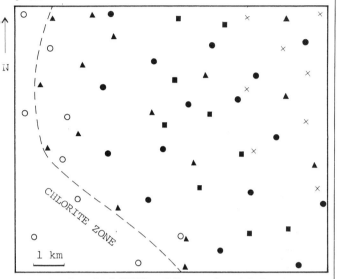

KEY TO MINERALS WHICH MAY BE USED AS INDEX MINERALS

increasing metamorphic grade

○ chlorite
▲ biotite
● garnet
■ kyanite
× sillimanite

Fig. 5.19 Exercise in plotting boundaries between metamorphic zones.

This exercise, using a hypothetical area, will demonstrate how the system works. The map (Fig. 5.19) shows discoveries of some of the index minerals in a group of metamorphosed pelitic rocks, and the boundary between the first two zones has been drawn. You should try to draw the other three boundaries on a tracing of the map.

The application of these principles to the mapping of the Grampian Highlands has led to the construction of the map in Fig. 5.20. Barrow did not carry out all this work himself so his original zonal indices have been somewhat modified. Nonetheless, the zones of most of the region are usually known as 'Barrovian type zones', in his memory. The lowest grade zone is the Chlorite Zone and the highest the Sillimanite Zone.

The map also shows another series of zones, the 'Buchan type zones' in the north-eastern part of the region. The principle of establishing the zones here is similar, but the index minerals are quite different. Although the fabric and structure of the rocks is characteristic of regional metamorphism, the index minerals are more akin to those found in contact aureoles (Fig. 5.1). The two sets of zones clearly indicate metamorphism under at least two different geothermal gradients, being intermediate in the Barrovian zones and high in the Buchan zones. It has been estimated that the geothermal gradient in the region of the Barrovian type zones lay between 30° and $50^{\circ}C$ km^{-1} whilst in the Buchan region it was probably between 50° and $70^{\circ}Ckm^{-1}$

In general, the metamorphic zones of both areas are relatively independent of the very complex fold structures of the Grampians, so much of the metamorphism must have occurred after the main phases of deformation.

Undoubtedly, the above account presents an oversimplification of the story. We now know, for example, that the metamorphism did not happen all at once, but in many different phases, described by the Institute of Geological Sciences as 'several high tides' of metamorphism. Nevertheless, the basic concepts formulated over 70 years ago still hold good.

Answers to exercise on page 26

1 = b, 2 = c, 3 = a

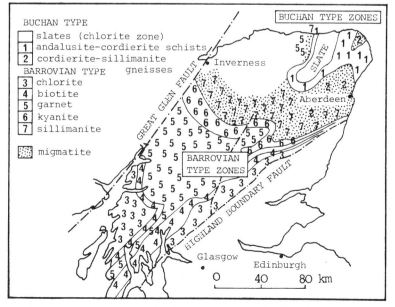

Fig. 5.20 Zones of regional metamorphism in the Grampian Highlands.

Metamorphism and Destructive Plate Margins

One rather distinctive group of metamorphic rocks, so far mentioned only briefly, are the so-called blueschists or glaucophane-bearing schists. Although they are not widespread and are usually only seen in post Palaeozoic orogenic areas, they seem to be very closely related to a particular geotectonic setting.

The blueschists derive their name from the minerals they contain. The diagnostic minerals of the assemblage are the blue amphibole glaucophane and lawsonite, a complex silicate. These are commonly accompanied by epidote, muscovite, albite, quartz and chlorite (Fig. 6.1). A study of the con-

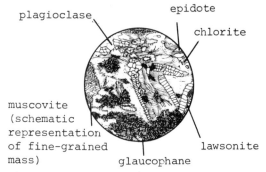

Fig. 6.1 Metamorphosed basalt, from Pic Marcel, near Guillestre, France.

ditions of formation of the characteristic minerals of the blueschists, especially lawsonite, shows that they must have developed under very high pressures of between 5×10^8 Pa and 10×10^8 Pa combined with temperatures as low as 150° C to 350° C (Fig.

6.2). This suggests that the rocks were buried to depths of over 20 km in areas of low heat flow. Blueschists can be derived from rocks of diverse composition such as basalts, impure sandstones or even cherts, but whatever the original rock, the geothermal gradient required is as low as 10° C km^{-1} in order to reproduce the correct conditions of temperature and pressure. Today, geothermal gradients as low as this are found only beneath ocean trenches (Table 6.1).

TABLE 6.1 GEOTHERMAL GRADIENTS AT THE PRESENT DAY

		$^\circ$C Km^{-1}
Tectonically active continental margins	(average value)	30
Ocean Basins	(average value)	20
Pre-Cambrian shields	(average value)	16
Guatemala Basin		10.8
Japan Trench		10.7
Marianas Trench		9.3

This type of rock can have rather confusing field relationships, and blueschists are often seen as small patches within low grade or even unmetamorphosed rocks. One such occurrence is in the Coast Ranges of California, where the Franciscan Formation is exposed. This formation consists of Jurassic and Cretaceous sediments which were probably deposited on the ocean floor. The rocks have been metamorphosed to blueschists

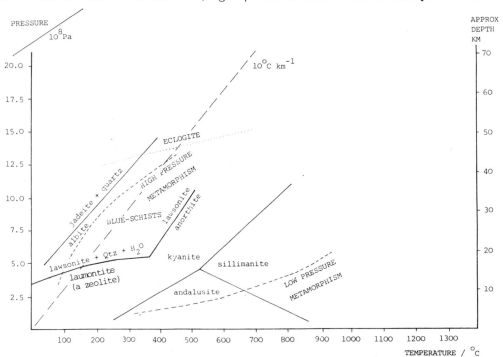

Fig. 6.2 Different temperature-pressure gradients in metamorphism.

which now occur as separate outcrops associated with serpentinised peridotite. We have already seen that for blueschists to form, the parent rock must have been subjected to very high pressures, although temperatures remained relatively low. It has been suggested that the downward movement of a lithospheric plate at a destructive margin would carry sediments to a depth where they would encounter the necessary pressures. The present position of the blueschists in the Coast Ranges is attributed to the continuing collision of the North American and East Pacific plates.

Confirmation of the association of blue-schists with destructive plate margins is found in the Japanese island arc. Of particular interest here is the fact that the blueschists occur in what are known as paired metamorphic belts (Fig. 6.3). The

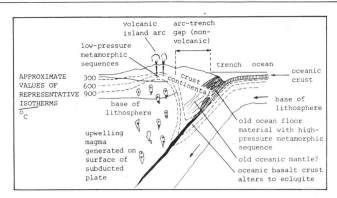

Fig. 6.4 Schematic model of lithosphere beneath an island arc. The large arrows show the motion of the oceanic lithosphere relative to the continental lithosphere on the left. The close spaced isotherms in continental crust beneath the volcanic part of the island arc indicate a high geothermal gradient. From Myashiro (1973) with sketch isotherms added.

The blueschists are formed as the oceanic plate is driven deeper into the mantle. Enormous rises in pressure will accompany this movement, but heat will only be conducted slowly so that the temperature will rise more gradually. This will give the low temperatures and high pressures needed for the formation of the blueschists.

By contrast, on the continental side of the island arc, the mainly pelitic rocks will be in a lower pressure regime where the heat flow is high due to rising magma and where andalusite and sillimanite will form (Fig. 6.4).

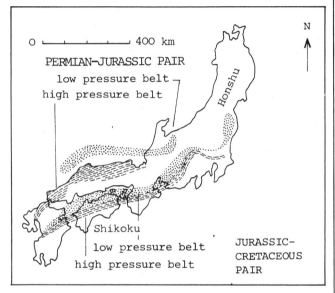

Fig. 6.3 Two pairs of regional metamorphic belts in Japan.

schists, developed from igneous rocks of basic composition, occur on the oceanic side of much of the island arc, indicating a high pressure regime. This is the so-called High Pressure Belt. Associated with the blueschists are eclogites, which are rather striking rocks of high density, containing green jadeite, rutile and red garnets. The eclogites are produced at relatively low temperatures but even higher pressures than the blueschists and are thought to result from the transformation of oceanic crust as it is forced down into the mantle along the subduction zone.

On the continental side of the island arc there is a belt of metamorphic rocks formed under conditions of a high geothermal gradient but low pressure. Hence it is known as the Low Pressure Belt. The mineral assemblage shows a great contrast with the High Pressure Belt, andalusite and silli-manite being the characteristic minerals developed from pelitic rocks. As in the High Pressure Belt, there is considerable variation in the grade of metamorphism. Figure 6.4 illustrates the possible method of formation of these paired metamorphic belts.

This pattern of paired metamorphic belts is also seen in other areas of present-day sub-duction, particularly around the circum-Pacific belt such as in New Zealand and Chile. It would be very satisfying if a similar pattern could be found in older orogenic areas such as the Alps or even older areas such as the Caledonides. However, in these areas the pattern is by no means as clear. In the Alps, for instance, the blue-schists are associated with metamorphic assemblages generated at intermediate rather than at low pressures. In the Caledonian chain, only small outcrops of blueschists have been found and few satisfactory deductions can be made. It has been suggest-ed that the high-pressure belts are present, but they are buried under younger sediments. As in the Alps, medium-pressure belts containing kyanite and sillimanite are found rather than low-pressure belts with andalusite and sillimanite (Fig. 6.2).

It has been suggested that one reason why the blueschists are absent from the older orogenic areas is that continental collision will have resulted in the remelting and hence obliteration of these rocks. The Franciscan Formation of California exhibits blueschists on the oceanic side of the orogenic belt, as in Japan. However, when continental collision eventually happens in California those blue-schists will almost certainly be reworked and the pattern obscured as has happened in the Alpine and Caledonian orogenic areas.

7
Metamorphism and the Rock Cycle

One of the greatest contributions made by
the late 18th century geologist, James
Hutton, was his realisation of the cyclic
nature of many geological processes. All his
geological research, and indeed all of ours,
has brought us no nearer to discovering the
'primeval rock' of the earth's crust, for
the simple reason that all the rocks we see
are recycled from even earlier material. We
have so far implied that the highest grade
metamorphic rocks represent an end-point,
but is there a sense in which they really
mark the beginning of the next cycle of
geological activity?

Weathering and erosion of metamorphic rocks

Once exposed to the atmosphere, our superb
crystalline metamorphic rocks fall prey to
the slow, but inexorable processes of
weathering and erosion. Just as in igneous
rocks, the constituent minerals are stable
at the temperatures and pressures at which
they are formed, but by the time they become
exposed at the surface they have been thrown
out of equilibrium and are more liable to
chemical attack. Physical processes, such
as freeze/thaw and temperature variations
take advantage of planes of foliation in
slates and schists, and breakdown occurs.
Gneisses are usually the most resistant of
the metamorphic rocks, but in time even they
succumb and decay to clay minerals and other
materials which are carried off to form new
sediments.

Melting of metamorphic rocks

Although we have regarded metamorphic rocks
as having been formed in the solid state, we
have also indicated that, in the most extreme
cases, some gneisses contain bands of
minerals whose random texture resembles that
of an igneous rock rather than a metamorphic
one. Such rocks are known as migmatites and
an example is shown in Fig. 7.1. This shows
clearly the contrast between the gneissose

*Fig. 7.1 Migmatite from the Lewisian, North
West Highlands.*

foliation in the lighter bands and the
granular texture in the intervening grey
patches. The lighter bands consist largely
of feldspars, micas and ferromagnesian
minerals, whilst the grey patches are mainly
quartz and feldspar.

The name migmatite means 'mixed rock' and
there is still much controversy regarding
their origin. Most geologists believe that
they can be produced by the partial melting
of gneiss, with very few additions from
external sources. It is also generally
accepted that the continued action of the
processes that produced migmatites may result in the formation of a granitic magma
(see the Unit *Igneous Petrology*). Because
of its relatively low density, such fluid
may then rise and become injected into higher
levels of the crust where it crystallises to
form a granite pluton. In due course, the
granite would be exposed to erosion and
would thus become the source rock for a new
phase of sedimentation. In other words,
there is a continuum between metamorphism,
igneous activity and sedimentation.

From observations such as these the concept
of the Rock Cycle has been derived, an
idealised form of which is shown in Fig. 7.2.
Much of this is in accord with the ideas
formulated by James Hutton, although in the
eighteenth century, subduction zones had not
been recognised and virtually nothing was
known about the earth's mantle. The reality
of the near surface processes are, however,
easily demonstrated, and in Hutton's day it
was quite reasonable to declare that there
was 'no prospect of an end' to geological
phenomena. Now, however, it is becoming
increasingly apparent that the Earth is
slowly evolving and that the internal
processes involved in the Rock Cycle are
never repeated exactly more than once. The
virtual absence of blueschists from strata
older than the Mezosoic (p. 30) may perhaps
be an example. Also, modern research has
shown that, although many of the world's
granites are derived from partial melting,
and hence, recycling, of parts of the continental crust, others originate as magmas
rising directly from the mantle. In this
case, the evidence has only recently come to
light during studies of the distribution of
isotopes of the relatively uncommon element,
strontium.

The time involved in such a large-scale
evolution of the earth is so enormous that
for all pratical purposes modern geologists
can see 'no prospect of an end', either,
but it will be fun to speculate about the
ultimate form of the earth, as further
geological discoveries give us a clearer
picture of its likely development!

Further Reading

Harker, A.
Metamorphism. Methuen, revised 1950. The 'classic' text on metamorphism (first published in 1932) based largely on detailed petrographic descriptions of metamorphic rocks. The book is liberally illustrated with drawings of thin sections.

Mason, R.
Petrology of Metamorphic Rocks. George Allen and Unwin, 1978. The book is intended for undergraduates but is well illustrated and is written in such a lucid style that students at GCE 'A' level will find many sections of it readable and helpful.

The Open University
Internal Processes, S23 Block 4. Open University Press, 1972. Reprinted 1978. The book contains a useful outline of metamorphic processes and the reader is encouraged to check what has been learned through a series of 'self-assessment' questions.

Read, H.H. and Watson, J.
Introduction to Geology Vol. 1. Macmillan 1975. A comprehensive text with a detailed section on metamorphism, intended for undergraduates and others.

References

Barrow, G.
'On the geology of lower Deeside and the Southern Highlands Border'. *Proceedings of the Geological Association Vol. 23*, pp.274-90, 1912.

Hutton, J.
Theory of the Earth. 1795.

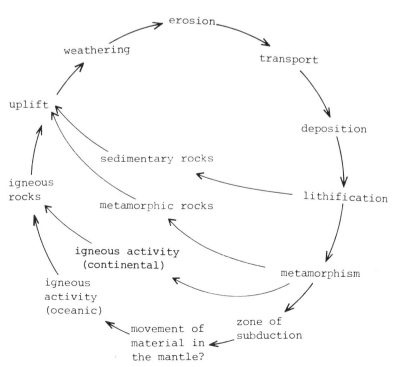

Fig. 7.2 The rock cycle.